Merry Christmas and *Best Wishes*
for Project Future's
success in '91.
Mary Ann K.

Omnicraft
incorporated

512 West Ireland Road, P.O. Box 2527
South Bend, Indiana 46680
Telephone: 219-291-7222

TREASURES OF THE TIDE

NWF BOOKS

NATIONAL WILDLIFE FEDERATION

Library of Congress CIP data: page 174

PREFACE

First, a mighty roar and thud. Then, with a swoosh, glistening foam swirls in retreat, leaving the sand as smooth as velvet. Somehow, the ocean's endless pounding against the shore is the most soothing sound I know. Whether you enjoy Florida's coral reefs or the Pacific's rocky coast, you can count on hearing the steady sound of waves.

But, because of increasing commercial and residential development, we cannot count on a secure future for our coastal treasures. This is why the National Wildlife Federation works diligently to protect the rich diversity of our coastal resources. Two areas of special concern are coastal barriers and wetlands.

South Padre Island in the Gulf of Mexico off the Texas coast is an example of a coastal barrier we can't afford to lose to development. It offers refuge for endangered sea turtles and scores of other wildlife, and it stands as a hurricane buffer for the Texas mainland.

Throughout the United States, such coastal barriers provide habitat for more than 20 endangered species and help maintain many kinds of shellfish and finned fish. The Federation is urging Congress to expand the Coastal Barrier Resources System. Once a coastal barrier is in the system, no federal subsidies can be used for its development.

Marshy wetlands along the southeastern Atlantic and Gulf coasts are essential breeding grounds for a multitude of wildlife, and they provide invaluable natural waste-treatment. But such wetlands are rapidly vanishing. Today, less than half of the United States' original wetlands still remain. Along with educating the public about the value of wetlands, the Federation is urging Congress to enact comprehensive legislation to safeguard these coastal areas. So when you listen to the soothing sound of waves pounding against a beach, or perhaps

the lap of muddy water against your boat in a Louisiana marsh, remember how urgently all of these coastal treasures need to be protected.

Jay D. Hair

President, National Wildlife Federation

CONTENTS

ATLANTIC

GULF

PACIFIC

ATLANTIC

Sandpipers, Sapelo Island, Georgia

Navigating the Atlantic Coast from Maine to Florida, a sailor needs to pore over more than 270 nautical charts. A wistful chronicler of the coast can just summon his own uncounted charts of memory: sitting with tin pail and shovel at the edge of the tide, skimming rocks between the waves at Cape Cod, peering into a tidal pool in Maine, beachcombing in Florida, watching a sea turtle lay her eggs in Carolina sands.

Snapshots of the shore. We put them in albums, but mostly we keep them in our memories, happily summoned by that magic phrase of childhood: *Going to the beach*. The beach of memory is a hazy stretch of sand and sea. Yet understanding the sandy beach comes only when we can get a feel for the rhythms that produced it, the recurrent forces of winds and waves.

Waves are the builders of beaches, working with fossil shells, rocks, or whatever substance is at hand. Some rocks were created thousands of years ago when glaciers gouged out boulders. Some are created today when waves cut into coastal headlands or when rock-carrying rivers unburden themselves into the

Lured by beachcombing (left) and picturesque settings, visitors flock to the Atlantic's sandy shores. Thanks to radar, many lighthouses no longer are beacons for navigation. Yet if one of these romantic symbols is threatened (such as Cape Hatteras, right) people fight to save it.

sea. Churned in an abrasive mixture of tossing sea and tumbling pebbles, the large rocks break down, becoming smaller and smoother.

Even the word beach seems to have once referred to a shore strewn with sea-smoothed gravel. It was known in English as a shingle, perhaps borrowing from Norwegians who called it a *singel*. They fancied that the rocks clattering and clacking underfoot were *singel*—singing.

I once walked a shingle beach in Maine with my nephew, who was studying to become a geologist. As he described what we were seeing, I began to notice how the stones had cobbled that beach: big, irregularly shaped stones near the low-tide mark, smaller, more uniformly sized stones higher up the sloping beach.

"The waves carry stones as far as they can," he said. "When the wave is so weak it can't carry stones, it drops them." We crouched near a stretch of coarse sand in the center of the beach. As we hunkered there, close to the interplay of stone and sea, I could hear the clacking of moving pebbles. To my ear, they didn't sing, though they clinked rhythmically, even musically.

A shingle beach may be a beach to a geologist. But to thousands of pail-toting youngsters, the essential ingredient of a *real* beach is sand.

And sand comes in wonderfully variable kinds and colors. On New York's Fire Island, sand often looks pink, created by red granules of garnet mixed with grains of white quartz.

In Florida, microscopic, densely packed seashells form the hard white coquina sands of Daytona Beach. This sand is created from the shells of ancient coquina clams, cemented by a carbonate ooze. In 1903, after drivers of new-fangled automobiles discovered the joys of speeding along this hard-packed surface, racing began at Daytona. The racers needed wide open spaces for their fast but unreliable cars, and they found a lack of things to hit on the 500-foot-wide, 23-mile-long beach. Racing has since been moved to Daytona's Speedway, although cars can still use the beach at legal speeds.

In Maryland I have found the tiny black teeth of sharks sprinkled through the sand, the stubborn survivors of grinding that wears away all but the hardest substance. That is why the most prevalent component in most sand is usually silica, the unyielding bits of quartz that put the rasp in sandpaper.

To a stroller on a beach, sand is sand. But to a geologist, a grain of sand is no smaller than 1/16th of a millimeter. (Smaller ones are on their way toward silthood.) And a grain cannot be bigger than 2 millimeters, about the size of this o. (Beyond is gravel and stone.)

You can stand on wet fine sand; your feet sink into coarse sand. A scuba diver discovered this one day, she told me, when she prepared to enter the ocean at an unfamiliar beach. Accustomed to the fine sand of her home beach, she foundered as a friend helped her with her weight belt and tank. She recalled, "I began sinking! Here I am planning to walk into the surf and my ankles are covered with wet sand."

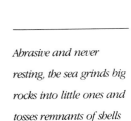

Sandpiper tracks crisscross the sand at a southern beach.

Abrasive and never resting, the sea grinds big rocks into little ones and tosses remnants of shells onto the shore.

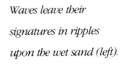

Waves leave their signatures in ripples upon the wet sand (left).

Next page: The sea acts as a medium for the wind's energy. The size of a wave depends on the depth of the sea and how fast, how long and how far the wind has been blowing.

Wherever the beach, it is being continually shaped and reshaped by wind and waves—perhaps 10,000 waves come ashore each day. But if you think of waves as rolling surges of water, think again. Waves are not water. Instead of a movement *of* the water they are more a movement *on* the water, energy bequeathed to the sea by the wind.

To standardize eyeball estimates of the force of wind, Francis Beaufort in 1805 rated the wind from calm (0) to "that which no canvas could withstand" (12). In 1955 the U.S. Weather Bureau expanded the Beaufort Scale from 13 (hurricane winds 83 to 92 miles per hour) to 17 (winds from 126 to 136 mph).

When wind touches the ocean, the power is transmitted to the sea. The greater that power, the higher the waves. As the bottom of a wave touches the sea floor near shore, part of the wave's energy is consumed by friction. The rest of the wave's energy continues shoreward. To an observer on the beach, it looks as if the wave has broken, with its frothy upper half—a breaker—crashing onto the sand.

As the weakening wave claws its way up the beach, gravity begins to pull the water back down the beach, leaving behind the sand that it carried and helped to create. When the dead wave nears the sea, a live wave washes over it in the roiling that makes the edge of the sea a glistening stage.

The beach's uppermost neighborhood is the area where the sea tosses its burden of wrack, an old English word for shipwreck. Bits of driftwood find a resting place here, along with shells, the bleached bones of fish, and bottles and other human litter. Amid the wrack on a beach in Maine, I once found a beachcomber's ultimate treasure: a note in a bottle!

Not believing my luck, I unscrewed the cap, reached in, and fished out a typed note from a Massachusetts man. The note was disappointing, for it consisted merely of his address and a request to report where the bottle washed up. I wrote him, telling him how thrilled I had been, how curious I was about the notes he had cast upon the waters. I was hoping for some fascinating correspondence, but I was never sent an answer, unless it was by a bottle still unfound.

There's a lesson here: The riches of the beach are best cherished as experiences and memories, not keepsakes. Ralph Waldo Emerson neatly captured the feeling:

> I wiped away the weeds and foam,
> I fetched my sea-born treasures home;
> But the poor, unsightly, noisome things
> Had left their beauty on the shore,
> With the sun and the sand and the wild
>   uproar.

The littoral or intertidal neighborhood is where the sea meets the land. Here birds are the most visible wildlife, with many species dining together with few signs of competition. Watch a mixed crowd—American oystercatchers, least sandpipers, brown pelicans, laughing gulls, least terns—flitting along a beach at low tide. Each has its own way of feeding.

The oystercatchers poke their chisel-like red bills deep into the wet sand in hopes of plucking out a clam and popping open its shell. Using its versatile bill as a tool, an oystercatcher also digs in the mud for fiddler crabs, pries limpets off rocks, or indulges in the activity that gave it its name—gorging on oysters.

While an oyster is agape, sifting for food in shallow water, the oystercatcher darts its bill into the oyster's shell and severs the oyster's shell-closing muscle. Then with sharp twists of its head, the bird wields its bill like an oyster knife, forcing the half shells apart and devouring

Roughly 10,000 waves a day wash ashore at Cape Cod (left) and other beaches. Along the Atlantic's wild edge are delights both natural and manmade. Overlooking Casco Bay in Maine are tall wooden towers, relics of World War II when spotters watched for German U-boats. Near the Kennedy Space Center on Florida's coast, eagles and peregrine falcons soar.

the meat. Sometimes the oyster manages to snap shut on the bird's bill. The oystercatcher responds by swinging its bill at a rock until the shell cracks.

The softly peeping sandpipers peck tiny holes in the sand, harvesting prey too small to be seen by a human observer. Chances are the meal consists of sandhoppers, or sand fleas.

Gulls like these tiny crustaceans too, and they get them out of the wet sand by dancing along the shallows. One scientist who watched the choreography speculated that the dancing "alarms the crustaceans, which rise to the water and scatter. As soon as they appear the bird stops its dance for a second or so, and, still remaining precisely in the same spot, snaps in the water at the swimming animals." To test his theory, he struck his fingers on the sand in an imitation of the dance. Sure enough, alarmed amphipods swarmed up and darted away.

Along the strand, pelicans waddle, looking comically awkward. Gazing toward the sea, they ignore the mingling terns and gulls. Then, one begins flapping its wings until it lurches into the sky. Suddenly it becomes graceful and flies off-shore, cruising a few feet above the waves.

Pelicans also fly high. They have been seen 70 feet in the air, diving and crashing into the sea, sometimes on their backs. The impact is softened by air sacs beneath the skin. Underwater, a pelican scoops fish into its fabulous bill pouch, then rotates its body to face into the wind for takeoff when it reaches the surface. After a pause to drain its pouch of gallons of seawater, it lifts its bill so that the fish slide, head-first, down its gullet.

Squawking maniacal cries, the laughing gulls dart into the surf, dipping to catch small fish. Or they go aloft long enough to fly to the high-tide line, where they scrounge for rotting fish or edible human litter.

All the oystercatcher needs is a slight gap in the shell of an oyster or other bivalve. Then it inserts its long, flat bill and severs the muscle that holds the halves together, enabling it to eat the meat inside. Oystercatchers are commonly seen along America's Atlantic shore.

If a cockle shell does not immediately yield the protected morsel inside, a herring gull (above) may smash it against rocks to shatter its shell. To feed its young, a royal tern in South Carolina (below, right) delivers a fish.

ghost crab. It moved swiftly on a crab's version of tip-toes, a pair of black eyes on stalks sticking out from its body-head, a pair of huge claws held out in a boxer's pose.

Shells forgotten, we watched in silence as the crab hurried down to the sea, leaving a sandy wake of scratchy prints. Later we learned that the ghost crab spends much of the day in its upper-beach burrow, where, sheltered from heat and sun, it lives on oxygen extracted from seawater that it keeps in gill chambers. It occasionally makes sea runs to replenish its life-giving water supply, or to forage for food.

In its burrow, a ghost crab may dig down as deep as four feet to reach moist sand. Then the crab burrows horizontally to make a tunnel. Sometimes at the end of the tunnel the crab digs a second opening to the surface, perhaps an escape route or a ventilation shaft.

After the crab brings food into the burrow, it begins an elaborate exercise in concealing its hideout. The crab gathers sand around the opening and enters. Only the tips of its walking legs stick out as the crab tamps down the sand. When it pulls the rest of itself in, the opening is

Terns fly closer in, black-capped heads down as they search for small fish or insects. John James Audubon, entranced as he watched terns feeding in a swamp, singled out one: "Up rises the little thing with the shrimp in its bill, and again down it plunges, and its movements are so light and graceful that you look on with pleasure and are in no haste to depart."

The herring gull, the most commonly seen gull along most of the Atlantic Coast, can open clams by dropping them on rocks. The gull plummets with the clam, landing near it to keep other gulls away. If there is no good shell-breaking place near the shore, the gull looks for one inland. Shattered clam shells have been found on roads a half-mile from the Atlantic.

Turn away from the shore and walk up to the high-tide line to find another creature that lives in peril of seabirds, the ghost crab. I first saw these aptly named creatures at twilight on a Florida beach. My children and I were using the last minutes of daylight to search for shells when a sudden movement caused me to turn my eye. Near a nickel-sized hole in the sand scuttled a

almost invisible. It remains closed until the crab's next sortie to the sea.

Ghost crabs usually live in colonies, but a crab never enters another's burrow. How does a crab know a burrow is occupied? Probably through signaling sounds made by rubbing the side of its big claw against three notches on its shell, a kind of three-note keyboard. One scientist listened to what happened when he forced one crab into the burrow of another. "The intruder," he wrote, "shows the strongest reluctance to enter and will take all the risks of open flight rather than do so.... When the rightful owner discovers the intruder, he utters a few broken tones of remonstrance, on hearing which the intruder, if permitted, will at once leave the burrow."

Such quiet dramas are played out thousands of times along the sandy beaches of the Atlantic. The beaches are the backdrop for day-to-day struggles and, for many of us, they stand as symbols of launching, of dawning. Perhaps here we see, as our forefathers did, a chance for a new beginning.

Stand on the sandy shore of Bulls Island, off the coast near Charleston, South Carolina, and turn your back to the sea. Rippling sea oats and other grasses dot the sand; rising behind the grasses is a tangle of wax myrtle. Beyond is a dark forest of magnolias, pines, palmettos, and live oaks, bent against the wind.

Brown pelicans dive bomb into the water to catch fish, but must stay alert as they emerge, lest hungry gulls steal their hard-earned catch.

Ghost crabs must make periodic runs from their burrows in the sand back to the water to wet their gills. On summer nights, they search for food, retreating by day to burrows (right) that protect them from predators and the heat. They often spend early morning hours repairing their burrows and stocking them with food.

The appearance of this wild shore has not changed since the time of Stephen Bull, who arrived here with other English settlers in 1670 and gave his name to this island. What he saw is there today. Live oaks still stand, their gnarled limbs draped in Spanish moss; seabirds still strut along the shore. Bull and the colonists who followed found vast woodlands thronged with "Turkeys, Quails, Curlues, Plovers, Teile, Herons ... Swans, Geese, Cranes, Duck and Mallard, and innumerable other water-fowl, whose names we know not. ..."

Today wildlife may not be so abundant, yet Bulls Island stands as a prime example of the human stewardship now bestowed on animal habitats. Bulls Island is part of Cape Romain National Wildlife Refuge, one of the Atlantic Coast's havens for birds and beasts. From Roosevelt Campobello International Park on the U.S.-Canadian border to Everglades National Park at the tip of Florida, sanctuaries—parks, national seashores, refuges—preserve wildlife along the ocean's edge.

Wildlife often struggles on that edge. In 1989, Hurricane Hugo ripped through Cape Romain. Towering, wind-lashed waves surged across Bulls Island, submerging the low-lying island and demolishing the new visitor center.

The concern for wildlife focused on Bulls Island, where red wolves had been reintroduced in an attempt to bring back a species that had once ranged as far west as Texas. Slaughtered for decades as varmints, the red wolves were dying off as a species in the wild.

In 1988, red wolves were released on Bulls by the U.S. Fish and Wildlife Service. Life in the wild has not been easy, for that same year an alligator killed the female of a pair. The male, which had shared in the feeding of the two pups, now worked doubly hard. In 1989, a new female wolf was put on Bulls and she and the male produced pups. Refuge workers later discovered an 11-foot alligator feeding on the female wolf. Again the male became a devoted single parent. Then, hardly a month after the second female's death, came Hugo. After the hurricane washed across Bulls, the adult male was found dead, but four pups survived.

On South Carolina's
Capers Island (above), a
weatherbeaten live oak
endures salt spray and
blowing sand. A few
miles south, a deer at
Hunting Island (right)
seems to find the sea as
refreshing as do the
humans who visit by day.

Numerous wild animals did perish during the hurricane—about 35 percent of those on the refuge, estimated manager George Garris. The damage might have been worse except most of the pelicans had already fledged and migrated south, and 90 percent of the loggerhead turtle eggs had already hatched.

No matter how devastating the storm, a host of animals live on: the alligators and the deer, the golden silk spiders and the yellow-bellied ribbon snakes. So do the herons and the egrets, the tanagers and the finches.

Pause along the mainland shore at low tide and watch a few of them. A great egret strides around, stabbing the water with its long yellow bill—stride, *stab,* stride, *stab*—in ceaseless rhythm, as silvery fish frantically leap and splash. A black skimmer swoops through the shallows, open bill cutting the water like scissors.

Romain—22 miles of bay, islands, shore-line, and river-laced marsh—is one of four wildlife havens that make up the Atlantic Coast's longest contiguous stretch of protected shore:

from a state sanctuary on South Carolina's North Island to state-owned land on Capers Island near Charleston. Included in this strip are many natural shields that protect the mainland from the full force of storms. They are the barrier islands, among them Bulls Island, that lie parallel to the coast, deflecting waves and winds.

Barrier islands are basically sand, constantly sculpted by wind and sea, often changing form or position along the Atlantic Coast. No such barrier islands shelter the United States' Pacific Coast, a shoreline that drops off steeply from mountain ranges rather than sloping off gradually from a coastal plain. The sands of Pacific Coast beaches come primarily from sediment carried into the ocean by rivers.

Occasionally the wind and waves of the Atlantic rearrange the landscape in startling fashion. In 1870, when North Carolina's Cape Hatteras lighthouse opened, it stood 1,500 feet from the sea. By 1989, engineers were working on ways to save it because so much beachfront had eroded that the lighthouse—the symbol of the well-loved Outer Banks—was now only 200 feet from the sea. Plans call for moving the lighthouse on rails half a mile inland.

The lighthouse guards Cape Hatteras National Seashore, a string of barrier islands that became the first coast given federal protection. The creation of the seashore in 1953 was inspired because shoreline development was threatening a part of the nation rich in history and natural beauty, the Outer Banks. Forests of live oak, cypress, and cedar had fallen to the axes of the earliest settlers. Stripped of stabilizing trees, the islands turned into domains of shifting sand.

At Nags Head, just north of the national seashore, is the colossus of the dunes, Jockey's Ridge, about 135 feet above sea level. At this popular spot for hang-gliding it is easy to see why Orville and Wilbur Wright chose nearby Kitty Hawk as the place where they soared into sky and history by powering their biplane into a gusty north wind.

About 140 miles north of Cape Hatteras stands another monument to the erosion that is the fate of barrier islands. In the late 1800s, the lighthouse at what is now Assateague Island National Seashore stood at the southern tip of what had been a long, thin peninsula running parallel to shore. Today it looms, a towering oddity, in the middle of a pine forest more than a mile from shore.

The northern end of the peninsula ceased to exist in 1933 when a hurricane punched an inlet through, making the peninsula an island cut off from the resort town of Ocean City, Maryland. The inlet might have eventually been closed by the normal southward course of seaborne sand. But two long jetties were built to keep the inlet open and capture the sand for Ocean City's beaches, which are shadowed by high rises and a boardwalk.

Barrier islands along the Atlantic Coast protect about two million acres of tidal wetlands and mile after mile of sandy beach. The barrier islands themselves—such as Martha's Vineyard off Massachusetts and Coney Island and Fire Island off New York—are often the beaches that lure us to sun and sand.

Yet unsullied stretches of coastline are rare between Delaware Bay and Cape Cod. And few barrier islands are wild like Bulls. Those big enough for development have become beach resorts, their geologic anatomy buried under boardwalks and cottages. Home builders flattened dunes and tore out sand-rooted plants. In futile efforts to stabilize the shore, developers, local governments and the Army Corps of Engineers built jetties that changed the natural shape of barrier beaches.

Now, geologists say, the Atlantic is rising and barrier islands are eroding and moving inland. As the wind and waves create a new profile, buildings and other human handiwork are

threatened. Martha's Vineyard's south shore is disappearing at the rate of about 10 feet a year; about 12 to 15 feet of Cape Hatteras vanish every year. "We know about 90 percent of the coast is eroding," says Stephen P. Leatherman, director of the Laboratory for Coastal Research at the University of Maryland.

Walking a barrier island in its wild state helps you appreciate nature's power—and the way creatures of the shore have adapted to life along the edge.

Cumberland Island, 16 miles long and one-half mile to three miles wide, is the largest of Georgia's 15 Sea Islands, which are strung along the coast from Savannah to the Florida border. Sea islands are barrier islands, but their core is solid bedrock, so they tend to be more stable, more capable of supporting maritime forests and a wealth of wildlife. Cumberland since 1972 has been a national seashore operated by the National Park Service and largely kept in its primitive state.

To get to Cumberland, I joined about 20 other visitors aboard a ferry in the little town of St. Marys. The ferry wound through the rippling marshes that touch both the solid shore and the

restless sea. But the trip was more than a passage from shore to island. In the middle of that passage, I could look at two worlds. Toward the mainland was the dazzle of morning sun on windshields and chrome, the silhouette of a sprawling mill, a sky tarnished by billows of smoke. Toward the island was an age-old seascape of blue-green water under a brilliant blue sky. A great egret rose from the marsh, its wings stroking with majestic slowness.

The ferry tied up at a dock; we walked to the small, cabin-style visitor center and got our orders from a ranger: "Take out what you brought in. There are no litter baskets here. Stay off the dunes. Don't try to feed any animals, especially the alligators." A pair of backpackers headed down one of the trails.

Visitors do not arrive on Cumberland casually, for space on the ferry—and even space on the island itself—is limited. Ecologists have determined that Cumberland, which is about a third larger than Manhattan Island, cannot sustain more than 300 persons at a time.

In a few minutes I was strolling a boardwalk across the dunes to the beach. For the moment I had 16 miles of beach to myself, and all I could see was an arching away of sand, sea, and sky. A string of pelicans skimmed the waves. Gulls were everywhere, waddling along the seam of sea and sand. The tide was just beginning to come in.

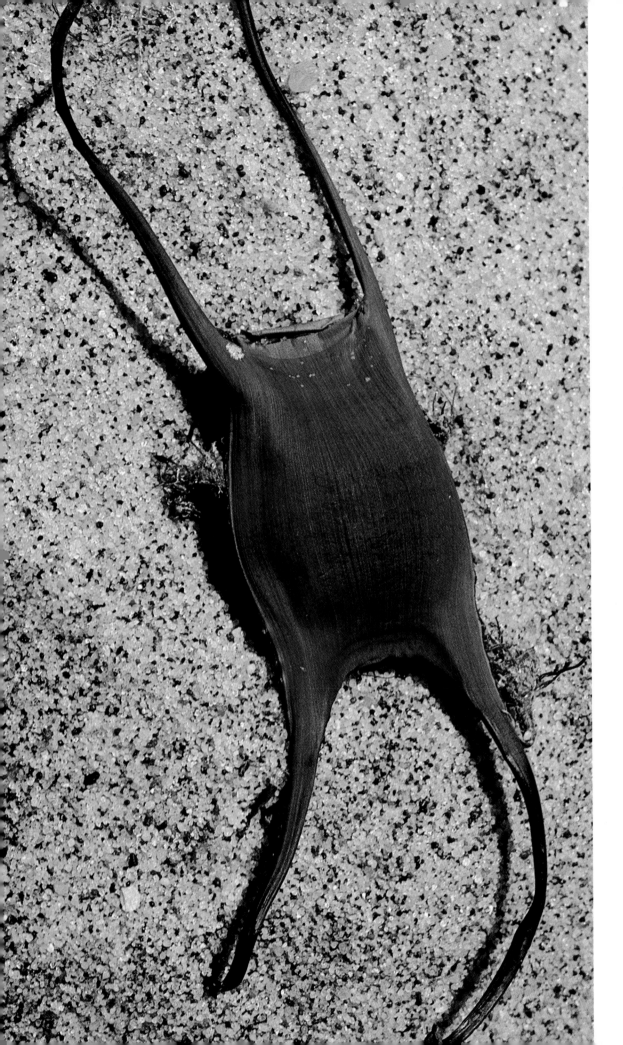

A mermaid's purse, the beautifully-named egg case of a skate, is among the treasures sometimes found in wrack on the beach. The tendrils on each end attached the packet to seaweed until the young skate hatched.

The sand was peppered with scattered holes so small that they could have been made with toothpicks. These marked the temporary abodes of tiny, rainbow-hued coquina clams. These fast-moving animals thrust themselves out of the wet sand as soon as they feel the break of a wave. Carried farther up the beach, they burrow into the sand, shooting out two barely visible siphons. One sucks in water for oxygen and wave-borne morsels, the other pumps out water and waste. Put a coquina back on the sand and it wiggles its foot, burrows and vanishes.

As each wave recedes, its last trickles form tiny V-shaped dams of sand grains on the landward side of the coquina holes. Then the incoming waves wash the Vs away. The rhythmic creation and erasure of the dams shows in miniature how waves affect a barrier island.

A walk along a pure beach like Cumberland's is a slow progress of stopping and stooping. I have explored dozens of beaches, each holding its surprises in scenes and smells and types of living things.

Off the Maryland-Virginia coast is a gem of the Atlantic shore, the 37-mile-long Assateague Island, which includes Chincoteague National Wildlife Refuge and Assateague Island National Seashore. The wilderness at Assateague holds mud flats and meadows, salt marshes and sandy beaches, and an abundance of fascinating wildlife. Strolling there at dawn one September, I spotted a mermaid's purse, a dogfish's skull, then a horseshoe crab.

The beautifully named mermaid's purse, about the size of a playing card, is the egg capsule of a skate, a kin of sharks. The mother deposits the dark brown, leathery capsule, anchoring it to a rock or shell with a sticky substance that coats one side of the capsule. Stiff, pointed horns project from all four corners, and

Without the stabilizing roots of sea oats, dunes erode quickly. An oft-told Mississippi legend recalls the man who harvested sea oats on the Isle of Caprice and sold them as floral decorations. Soon the entire island vanished.

I saw a sand collar—the eggs of a moon snail stuck together with sand—and admired how it had come to be: A moon snail rises from the wet sand, poising herself at a spot where just enough water gently laps around her, bearing grains of sand. They cling to an invisible film secreted by the snail. Slowly the grains become a wall that will guard the eggs the moon snail lays. After this quiet maternal drama, all that is left is the fragile, brittle sand collar.

sometimes the horns do the anchoring job by imbedding themselves in a muddy or sandy bottom. Firmly anchored, the flat packet becomes an incubator. After the young skate hatches, the case washes up on the beach.

In June I often saw the strings of egg cases of female whelks (aquatic snails). A whelk lays as many as 100 microscopic eggs in each of many parchmentlike egg capsules she produces in a special pore on her foot. The capsules hang on a long, twisted filament like wash on a clothesline. In time, hundreds of baby whelks emerge from the capsule, each in a tiny shell of its own.

Beyond the rim of sand are more surprises, and Assateague, like many a seashore, always dispenses discovery. My wife and I were watching a flock of laughing gulls one day when we saw—just beyond the surf—fins knifing through the water. For a moment I thought that the fins belonged to sharks. Then a body arched out of the sea and I could see the pointed snout and glistening shape of a bottlenose dolphin.

We stood there for long moments, watching the dolphin parade go by. They were hard to count—five? six? seven?—leaping and diving, appearing and disappearing like sprites.

On certain summer nights at Assateague, the ocean sparkles with tiny points of light. In the silent darkness, single-celled protozoans named *Noctiluca scintillans* ("shining light-of-the-night") produce luminescence by releasing special chemicals. Scientists cannot agree on

Lucky beachwalkers may see bottlenose dolphins knifing through the waves. This pair was spotted off the Florida coast.

why the shining-light magicians do this. Perhaps they light up to attract the opposite sex, or to scare off enemies. Standing there, entranced by their light, do we need a reason?

Yet my candidate for the Atlantic seaside's most fascinating animal is the horseshoe crab. Usually, all a beachcomber finds is something that looks like a stiff, hinged brown shell with a long, hard tail sticking out of it. Turn it over and you see a mass of mysterious body parts, all coated in a shiny material.

Actually, the shell is not a shell. The shiny material is essentially a cast-off skin. And the horseshoe crab is not really a crab. It is a form of marine spider, a relic of ages past.

The horseshoe crab slogs along the sea bottom mud by arching its body and pushing with its rodlike tail and rear pair of legs. Those legs also grab and crush food, usually worms and mollusks. The crushed food is picked up by the other four pairs of legs, which grip the pieces and pass them into the animal's mouth.

Each spring crowds of male and female horseshoe crabs gather on sandy shores to mate. Each female deposits 200 to 300 fertilized eggs in the wet sand, just below the high-tide line. Several weeks later, if the eggs have survived predation by shorebirds, they open and the young emerge. They usually spend the winter in mud flats feeding on tinier organisms.

On a secluded beach one May, I saw my first living horseshoe crabs. They were youngsters, scurrying around the shallows in their first molt. As they grew up, they would go through as many as 15 more molts, outgrowing and shedding their shells until they matured.

Because the crabs shed their skins so often, it's not unusual to find the cast-off remnants amid the dried-up seaweed, broken shells, and silvery driftwood.

Here, at the reach of high tide, is where the fine sand begins to pile up, forming the fore-dune line, the first defense against the sea. Here, too, is the beach's first vegetation, the sand-hugging, round-leafed beach pennywort and the golden sea oats, strong and steadfast. Long stems, stirred by the wind, scribe their endless calligraphy on the sand. There is beauty here. And protection. Sea oats send out rhizomes, underground plant stems that form a fibrous network knitting the sandy slopes together. Torn out and scattered by a storm, pieces of sea oat will stubbornly take root again and begin knitting at new sites.

On Cumberland Island I strolled down the dune-protecting boardwalk and looked upon the long arc of the interdune meadow. This sandy sweep of grasses and brush is bounded on one side by the foredune and on the other by the rear dunes, which rise about 45 feet.

A man leaned on the rail of the boardwalk, his gaze on a clump of green 50 yards away. He pointed and said, "That's the third one I've seen." For a moment I saw nothing. Then in some shrubby brush I spotted a patch of brown: the half-hidden head of a white-tailed deer.

Abundant along the Atlantic Coast, horseshoe crabs (left) come out in the spring to lay eggs in depressions scooped out in the sand. Hungry birds wait expectantly (above) for a chance to snatch and eat the eggs.

Cumberland's live-oak forest, dotted by alligator-prowled ponds and surprises such as armadillos (right), begins to thin out on the mainland side. Beyond are the salt marshes, a world of hidden wildlife and specialized plants.

I continued down the wooden stairs into the island's inner dominion, the marine forest. On Cumberland and other barrier islands of the southern Atlantic, the forest can be dominated by live oaks; on islands to the north, the principal trees are pines. On most islands there is also a band of thickets, where small trees and shrubs —and often virulent patches of poison ivy — take root.

In a single step I went from the radiance of Cumberland's sun-struck beach to the deep shadows of the live-oak forest. Thick, twisted limbs, shaggy with vines and Spanish moss, reached out and up, forming a dark, knotted roof over a setting for some Gothic fairy tale. Knee-high around the trees, like squat sentinels, were saw palmettos with long, spiky blades. Suddenly there was a rustle in the leaf litter. And

there beside the path appeared an unexpected walk-on to the Gothic tale: an armadillo.

A ranger later talked to me about this newcomer, the only mammal with a bony shell. "You should have been here in February, when the young are born," she said. "You look down and see a mother walking along with four *identical* little ones." The nine-banded armadillo gives birth to quadruplets produced by the division of one fertilized egg. Latin American animals introduced into Florida in the 1920s, armadillos were noticed on Cumberland in 1973 and have settled in as permanent residents.

Unlike the armadillos that noisily amble along the leaf litter, the stealthy bobcats are rarely seen or heard. Bobcats were common on Cumberland until the early 20th century, when disease and hunters wiped them out.

In 1988, wildlife researchers put 14 bobcats on the island, all wearing radio collars to help researchers keep track of them. One bobcat surprised the trackers when its radio beeps showed that it was swimming back to the mainland. The rest of the bobcats stayed, and all but one survived into the following year. A generation of island-born bobcats is now thriving. "We've seen ten kittens and think there are three more," a researcher said.

Cumberland is also one of many places along the coast, from Florida to North Carolina, where loggerhead sea turtles lay their eggs. Female loggerhead turtles, which weigh 200 to 300 pounds, return annually to the beaches where they themselves hatched. Not long after a summer sunset, a ponderous shadow emerges from the sea and lumbers up the beach to the high-tide line. Her huge carapace knobbed by barnacles, her rheumy eyes blinking, she heads for a clear patch of sand. By shifting her body from side to side, she settles into a shallow pit.

Scooping alternately with her rear flippers, she flings sand onto her carapace, slowly hiding herself as she digs an urn-shaped hole beneath her. Then she begins to lay her eggs, which are soft and about the size of Ping-Pong balls. After laying more than 100 eggs, she covers the nest, swishing around to wipe out all traces of the nest. Task completed, she heads back to the sea. She seems to be crying, but the tears are a secretion that protects her eyes from the tossed sand.

In about two months, the eggs will crack open. The hatchlings will stay in the shells a few days, absorbing their yolk sacs and waiting for their soft, curled carapaces to straighten and harden. Then, struggling together in a mass escape, they will dig themselves out and crawl to the ocean, males to spend their lives at sea, fe-

males to live and mate there, returning to the land only to lay eggs.

Deep in the sand, the hatchlings seem to sense the safest time for their breakout, which usually occurs after dark. Night or day, however, many of the tiny turtles will not reach the sea. Seabirds, crabs, and raccoons feast on the newly emerged turtles as they creep toward the surf. About one in 10,000 hatchlings lives to maturity. Since the loggerheads are a threatened species, such odds seem all the more daunting.

Yet the rhythm of life—the silent dramas of turtles and crabs, the urgent clatter of gulls and sandpipers—gives the sandy beach part of its character. Along America's sandy shoreline lives this crowded community—governed by the miracle of the tides, adapting to the forces of wind and water, delighting human onlookers who take the time to observe. 🐚

Scientists believe female loggerhead turtles (left) return to the beach where they were born to lay their eggs.

The Atlantic Coast, marsh-fringed and shielded by barrier islands for much of its length, offers a chance to see the differences between a sandy shore and a shore laced with tidal creeks and rivers. A sandy beach seems only touched by the sea. A marsh almost pulses with the sea, taking water and giving water as the tides come and go. Sidney Lanier captured this ecological reality of the marsh in his tribute to Georgia's Marshes of Glynn:

> *Ye marshes, how candid and*
> > *simple and nothing-withholding*
> > *and free*
> *Ye publish yourselves to the sky and offer*
> > *yourselves to the sea!*

The *simplicity* of a marsh is what first catches our eye: acres of swaying cord grass, each acre monotonously like the acres around it—golden in fall and winter, green in summer and

Shallow waterways and wide swaths of cord grass in a salt marsh (left) provide a rich source of food for many animals. Great egrets (right) stand motionless until they spot prey, then quickly stab it.

spring. Each day, with predictable regularity, the tide comes in and the tide goes out.

On the surface there seems to be nothing surprising about a marsh, except perhaps the smell. During high tide, we inhale the tangy salt of the sea; during low tide, we smell the dank odor of decay. The complexity of the "candid and simple" marsh is revealed in this rhythm of tang and decay, birth and death.

At the sea edge of a typical marsh, the fresh water of rivers and the salt water of the sea meet. Bacteria of seemingly infinite variety decompose dead plants and animals, creating an aquatic version of a gardener's compost pile. The food flows in twice a day on the high tide, sustaining shrimp, oysters, periwinkles, crabs, fish, and numerous other animals. Veteran observer William A. Niering wrote ". . . a properly

protected marsh cannot be depleted. It continually produces, like the legendary pitcher of wine that is never emptied."

The water's salinity would kill the typical land plant, but cord grass has adapted to salt water and reigns here in triumph. Cord grass stores a higher concentration of salt in its cells than land plants do, so the balance with outside salt water stays even. Sea water does enter cord grass roots, but membranes in the roots bar most of the salt. The remaining salt is plucked out of the plant's system by special glands that pass the salt to pores on the stems. At low tide you can often see these tiny crystals of salt glittering in the sun.

Cord grass roots, like most living things, require oxygen to breathe, yet the marsh's dense layer of mud contains little of it. The solution? A set of hollow "breathing" tubes extending from cord grass roots to openings in its leaves. When high tide comes in, the leaf openings close and the pipelines remain dry.

At low-tide level, a tall, coarse cord grass (*Spartina alterniflora*) rules the marsh. At the higher tide level, it's *Spartina patens*, a shorter, finer grass that grows no taller than two feet. The slender stems of *S. patens* tend to sway against their neighbors, sometimes bending the whole mass in a swirled cowlick pattern. As the densely clustered *S. patens* dies, the dead grass turns into a protective mat. The taller *S. alterniflora* stems tend to break off and wash away, leaving a stubble of broken stems around the marsh.

For animals, adapting to the salt marsh simply means avoiding the changes the tides bring. Birds can fly; mollusks, worms and most crustaceans burrow into the mud when the tide goes out. Fiddler crabs, perhaps the marsh's best-known inhabitants, flow at low tide like a dark shadow across the mud flats. Anyone who has seen a fiddler ballet must have wondered what happens to the performers when their stage disappears under high tide.

The fiddlers escape beneath the stage, digging burrows a foot or more deep. When the tide comes in, they crawl into their holes. Wave-borne sand seals the holes, allowing the crabs to live on trapped oxygen. At low tide, they emerge and dine on microscopic animals.

When eating, a fiddler crab scoops up sand and mud with "spoon-finger" claws—two on a female, one on a male. Opposite the male's small claw is the large claw whose shape inspired the fiddler's name. Male fiddlers sometimes wave their large claws to warn off intruders, and sometimes to attract females.

The waving does indeed interest females. When a female responds to the male's semaphore message, the two enter the male's burrow to breed. The female attaches the fertilized eggs to her abdomen and carries them until one day, at dusk, she goes to the water's edge and gyrates, catapulting larvae out into the sea. The tiny crabs feed on the invisible animals and plants that throng the sea.

Camouflaged in the muddy marsh bottom, the little crab emerges from its first covering and waits for the completion of its second. After five such moltings, each about a week apart, the crab looks like a shrimp with the legs of a crab. It swims on the surface for nearly a month. Then it finds a place to hide and molts at least four more times, finally coming forth as the fiddler crab we see scuttling about the dark mud of the marsh.

A marsh also welcomes many visiting mammals, perhaps a resolute raccoon or a red fox skulking around in search of eggs. The resident mammals, rabbits and rats, are usually seen as furry brown flashes.

By rhythmically waving its large claw, a male fiddler crab intimidates his rivals and beckons females into its burrow, which stays cool and moist because of its location just below the high-tide mark.

Confronted by a predator, a marsh rabbit will either swim away or hide by remaining motionless in the water. "On touching them with a stick," naturalist John Bachman reported, "they seem unwilling to move until they perceived that they were observed, when they swam away with great celerity."

Marsh rice rats build a complex habitat above the tides. They weave grass nests that hang on stems, and from the nests they build runways to feeding platforms where they dine on seeds, plants, and whatever bits of protein flit by. The rats have a life expectancy of about a year because they themselves are a major food source for owls, hawks, and snakes.

Marshes attract several species of long-legged wading birds. You can count on seeing an egret dipping its bill in search of food, or you may see a tricolored heron, its outstretched wings cutting down glare to make hunting easier. But don't expect to see a bittern.

The well-camouflaged bittern has stripes running down its throat, breast, and belly. When a person approaches the bird, it freezes, stretching its neck and pointing its bill upward so that its stripes become vertical lines that blend with the cord grass. If a breeze stirs, the bird even sways with the grass.

The more visible birds of the marsh—the strutting, poking, gobbling egrets and herons—fly in to feed and then fly out. The secretive rails live there, building their nests on grass clumps scant inches above the high-tide line. The little birds slip through the grass so furtively that they seem to pass between the stems. To do that, you would have to be "thin as a rail," a phrase that early observers contributed to the language.

Once in a while at low tide, you can see a rail searching a mud flat for food. A favorite is fiddler crab. If the victim is a male, the rail disarms, or, rather, disclaws the crab before eating. The bird vigorously shakes the fiddler until his big claw falls off. Then the rail safely goes on with its meal.

Today we can only imagine how many rails once lived in tidal marshes from New England to Texas. John James Audubon reported he collected 72 dozen clapper rail eggs in one day in New Jersey. He was still well below the typical egg-hunter's daily haul of 100 dozen. The clapper rail survived such plundering, but it never again inhabited marshes in such large numbers.

Another familiar marsh resident, the terrapin, was also nearly killed off. The terrapins, often called diamondbacks for the designs on their carapaces, once were so abundant that they were fed to slaves several times a week. The slaughter of terrapins so reduced their numbers that they became a rarity; today they are protected in several areas.

At home in brackish or fresh marshes, the American bittern freezes in a skyward-pointing pose when threatened. Its striped feathers aid in the effective camouflage.

As the Atlantic Coast ribbons southward, hundreds of acres of marshes are interspersed with estuaries and rivers that drain into the sea. As the climate changes, so does the character of the marshes. In the maritime forest of Florida's Canaveral National Seashore, wild coffee trees and other tropical species appear amid the live oaks, portending the changes in the southern Florida coast. There mangrove trees with stilt-like roots replace the cord grasses that carpeted the marshes to the north.

Two national parks in Florida, Biscayne and Everglades, have such mangrove swamps. The swamp at Biscayne is small and surrounded by increasing development, but you can walk around the stand of mangroves and understand their role as replacements for cord grasses. Peering through the jumbled veil of prop roots, you can see how they trap the trees' fallen leaves and other plant debris.

The decaying leaves add protein to the nutrient soup that feeds tiny animals so obscure they bear only Latin names. The outgoing tide carries the nutrients to sea, feeding animals farther from shore. With the incoming tide comes a fresh cast of animals seeking food. Tiny creatures are not the only ones drawn to the mangroves. I have often seen lumbering manatees pop up for air, their homely faces turned toward the boat I was aboard. The endangered manatees are so docile that they often are oblivious to speeding boats, and many bear scars of wounds made by propeller blades.

The human encroachment that threatens the manatee also threatens the Everglades, 2,200 square miles of marshlands that shelter North America's most spectacular assemblage of birds. In a few minutes you can see great blue herons and egrets striding in the shallows, flocks of white ibis aloft, roseate spoonbills gracing the mangrove trees, and snake-necked anhingas perched on shore, wings outstretched to dry.

Such sights bring hordes of human visitors in winter, but on some steamy 90° July days, the only hordes around are mosquitoes. Then the Everglades' shores become perhaps the *least* populated stretch of Atlantic coastline. It is a somnolent counterpoint to a summer day 1,000 miles to the north, where cars edge along at 10 m.p.h. on Maryland's Bay Bridge, gateway to the shore for thousands living in Washington and Baltimore. Hundreds more people are setting sail from points along the Chesapeake Bay.

While traffic and crowds are bothersome, the most visible natural hazard of Chesapeake Bay may be the Portuguese man o' war, an unwelcome visitor that ranges over much of the Atlantic shoreline. The Portuguese man o' war (scornfully named by the British when the Portuguese had a third-class navy) is the most notorious member of the clan that also includes the jellyfish. Its sting, rarely fatal but always painful, is compared to a severe electric shock.

A man o' war is nasty perhaps, but wondrous too, for it is an assortment of creatures. The man o' war is actually a community, a collection of separate animals upon which scientists have bestowed the designation *persons*.

What a casual observer sees is a purple, blue, or pink balloonlike sail moving with the wind. But what is passing by are "persons," each with a specialty. The colony in the sinuous tentacles catches and ingests food; other groups focus on flotation or reproduction.

The eating group and the defensive group together form the stinging mechanism and the tentacles, which may reach 50 feet long. As the man o' war floats along, fish swim between the tentacles, triggering stinging cells. These shoot microscopic needles into the fish and inject a paralyzing poison. The tentacles then pass the

fish to the feeding group; specialists in digestion later take over, transforming the prey into nutrients for the entire community.

The man o' war does not seem to seek out prey. When a flotilla of the creatures appears off a swimmers' beach, the raid is most likely not deliberate. Going wherever the wind propels it, the man o' war may be washed ashore to die, a jellylike mass. But, as many a beachcomber has painfully learned, the stinging cells live on and can still inject poison.

The most common jellyfish of Chesapeake Bay is the pinkish-red sea nettle. Its tentacles, which may number up to 40, are equipped with a similar kind of stunner: a coiled tube that springs out, imbeds itself in the victim, and injects a poison.

The word Chesapeake, meaning "great shellfish bay," was given to the bay by the Algonquin Indians. A different account calls it "Mother of Waters" in another Indian language. Both are good names, and both are appropriate, for the bay has been nurturing people ever since Indians began tossing their oyster shells into great piles that archaeologists still are finding.

The Chesapeake Bay is nearly 200 miles long and four to 30 miles wide. But its shoreline,

Maryland's South River is among dozens of the tributaries pouring into the Chesapeake Bay (at top of photo). The bay remains a productive fishery in spite of threats to its ecosystem. The Portuguese man o' war (below) is a threat to swimmers in the bay due to its painful sting.

if all the tributaries of all its rivers are included, zigs and zags for about 8,000 miles, making it the largest estuary in North America.

The bay is relatively shallow. From Professor M.G. Wolman of Johns Hopkins University comes a way to visualize the shallowness: If the bay were scaled down to be as wide as this page, its depth would be about one-third the thickness of this page. Because it is so shallow, the bay is sensitive to the general sea-level rise along the Atlantic Coast.

The history of Chesapeake Bay provides dramatic records of the constant disappearance of shoreline. Captain John Smith, exploring the lands claimed by the Virginia Company in 1607, saw a shoreline that is 200 to 2,000 feet from today's bay coast. Sharpes Island, at the mouth of the Choptank River, encompassed 438 acres around the turn of the century. In 1914, when Sharpes was down to about 50 acres, an erosion

Using power-operated "patent tongs," an oysterman dumps his catch onto a culling board to sort it. Laws conserving the fishery require that undersized oysters be returned to the exact bed from which they were taken.

study predicted that it would vanish by 1951. It did, and now exists on charts as a shoal.

Fierce storms also carry away shoreline. One bay peninsula, hammered by a severe three-day storm, lost 40 to 60 feet of bank. Another stretch of bay shore suffered "shoreline retreat" ranging from one foot a year to 36 feet *per month* for one five-month period.

Along with natural causes, the onslaughts of civilization have been threatening the bay's storied abundance. The sage of Baltimore, H.L. Mencken, called Chesapeake Bay an "immense protein factory." The bay sustains about 2,700 plant and animal species, though pollution and silt are slowly killing the bay's riches.

The bay's oysters were described by early colonists as "very large and delicate in taste." They still are, but there are not as many of them as there once were. In the 1870s, about 17 million bushels of oysters were plucked from the bay each year; today the annual harvest is about three million—and it keeps dropping.

Oysters are still being harvested by crews who are known locally as watermen. Their ship is the skipjack, a rakish craft with her tall mast set well forward, a jib rigged from a bowsprit, and a triangular mainsail. The earliest skipjacks could be sailed by one man who, if he had to, could also operate the dredge or "drudge," a wooden bar (later a metal one) with iron teeth that was dragged along oyster beds. The teeth pulled up the oysters into a netting bag attached to the bar. During the 1880s there were 2,567 licensed, sail-powered drudge boats in the bay. Today there are about 30, and people wonder if a day will come when no more skipjacks work the bay.

Bay birdwatchers also worry about a day when waterfowl will no longer flock here. Redheads have long been in decline. Widgeons, swans, and canvasbacks are also showing up in lower numbers. Fishing bans on striped bass,

known in the bay as rockfish, and shad were imposed in recent years to protect spawning fish. The bans helped, though a 1989 report on the state of the bay said they are "at dangerously low levels."

The seasonal flow of seawater in and out of the bay governs the life of the Chesapeake's best-known inhabitants, blue crabs and rockfish. The warm, incoming sea of late summer and early fall signals the crabs and rockfish to swim from tributaries to the bay. In spring, responding to changes in temperature, the crabs and fish return to the upper reaches of the bay.

The crabs end their spring journey in marshy shallows spiked with the long, narrow leaves of eelgrass. "They go there to hide and to feed and to feel the rays of the warming sun. And think about such things associated with spring," wrote William W. Warner, the perceptive biographer of the blue crab.

But the grass—once an underwater prairie of half a million acres—has been vanishing. Its disappearance is both a symbol of the bay's troubles and a real loss that affects virtually every creature in the bay's ecosystem. To find out why the grasses are dying, scientists have been working to solve what one of them calls "an enormous scientific detective case."

They concluded that the grasses, though hurt by a flood of chemicals, were dying because of a combination of erosion and nutrients. Millions of tons of soil are carried into the bay by springtime flooding. The silt clouds the water, keeping sunlight from the grasses and stunting their growth. Also borne by the water are fertilizers draining from farmlands, along with nu-

Chesapeake Bay's oyster beds are worked by the only commercial sailboat fleet in the United States. Wooden-hulled skipjacks (above) sport bowsprits and elegant, raked masts.

Next page: A channel marker makes a handy site for these ospreys' untidy nest of sticks.

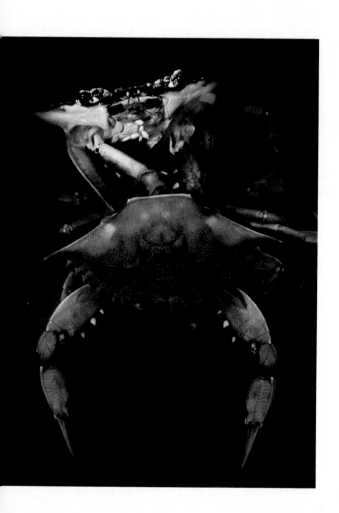

As it grows, the blue
crab repeatedly sheds its
shell (bottom in photo).
When gathered within
24 hours of molting,
the soft-shell crab is
a culinary delicacy.

trients produced by the breakdown of sewage at the more than 500 treatment plants around the bay. The nutrients feed microscopic plants that live on the grass stems. These tiny plants grow so profusely that they cover the stems, shutting them off from light.

The population pressure on this part of the coast will probably only get worse. Today about 10 million people live in the Tidewater, a maritime region with the Chesapeake Bay as its hub.

The attraction of the coast is nothing new, of course. Chesapeake is one of the many bays, estuaries, and other variations of the open beach that offered safe harbors for people wanting a place to settle and prosper. But by the late 19th century, the coast from New Jersey to Boston had become a habitat for the sharks of com-

merce, as well. Railroads and roads ran from the old port cities to the new cities, which burgeoned with more people, more development. Shoreline property became the most valuable real estate in the Northeast, and seaside resorts like Atlantic City flourished.

The coast was and is a generous provider of food and fun. Yet its fertile bays and marshes, ten times more fertile than a good wheatfield in Iowa, now have become threatened by the same human development that has surged along the Atlantic's sandy shores. Left to their own natural processes, the adaptable marshes and bays would have been well able to take care of themselves. Threatened by the hand of man, they stand in danger.

very summer my wife Scottie and I go to Maine and enjoy the splendors of its coastline: the sun lighting up a granite cliff, the sea sweeping across a boulder-strewn shore, a seal slithering up a pile of rocks bared by an outgoing tide. The vistas always include rocks, for rocks are the backbone of Maine's twisting coastline.

To understand the power of the sea and the perils along the Atlantic Coast, you need only look at the rocks—and what happened to the wooden sailing ships that passed this way in the 19th and early 20th century. Some of the world's largest wooden sailing ships were built in a shipyard in Bath, Maine. Of 45 ships built in the dying days of sail, all were lost, many on the rocky, fogbound coast. The toll begins with the *Eleanor A. Percy,* the largest schooner in the world when she was launched in 1900. She sank with all hands in 1919.

About 100 miles up the coast from Bath, Acadia National Park offers gentler revelations about the rocky shore. Most of the park sits within Mount Desert Island, which was part of the mainland until the glaciers melted thousands of years ago.

On dark, predawn summer mornings, visitors throng to 1,530-foot Cadillac Mountain, where flat granite rocks crown the

Rocky shores like the pinkish granite coast at Maine's Acadia National Park (left) offer the best opportunities to watch marine animals. Tidepools harbor a variety of hardy creatures while crannies on the rock face provide root holds for harebells (right) and other plants.

Arctic to North Carolina, from just below the low-tide line to waters 75 feet deep.

The sea anemone attaches its disklike foot to a rock and lives on the food swept into its mouth by its tentacles fluttering in the currents. Some species also slide along on the disk. The ghost anemone has 60 pale, shimmering tentacles around its mouth. When disturbed, it fires off a mine field—gossamer white strings that surround it. An intruder touching one of the strings is stung.

The frilled anemone, which gets its name from its fringe of about 1,000 tentacles, has a similar mine-field defense system. The frilled anemone can reproduce sexually or by taking a solitary stroll. When it is in this asexual mood, pieces of its single broad foot—technically, a pedal pad—adhere to surfaces that the frilled anemone passes over. This forsaken tissue then regenerates into new anemones.

Tidepoolers, hikers and rock scramblers are attracted to sea-sculpted caves (above), created when turbulent waves wear away softer strata of rocky headlands. The red sea anemone (lower right) uses its tentacles to poison its prey.

summit. Sitting on those chill rocks before first light and waiting for the sun to come up, they are among the first people in the continental United States to be touched by that day's dawn.

At a headland in the park, gnarled rocks climb from a misty sea. Here Bass Harbor Head Light has pierced the fog with a warning beacon since 1858. A trail winds through the firs that cling to the rocks. Framed by their branches are lobster boats and sailing ships.

Farther around the island's shore is Thunder Hole, which gets its name from the explosive sound of trapped air being released. When a wave surges into the hole, it compresses the air within. As the wave recedes, the outrushing air exits with a thunderous roar.

A quieter product of the waves is Anemone Cave, named for sea anemones, animals whose petal-like tentacles look like flowers that bear the same name. Because careless visitors trampled the area around Anemone Cave, its location has now been erased from park maps. But anemones can still be seen in other places. On the East Coast, anemones are found from the

Flashing red every four seconds, Bass Harbor Head Light (right) guides mariners around Maine's rocky Mt. Desert Island.

Acadia and other masterpieces of Maine are waystations for us as we enjoy our annual stay at an 80-acre island in Casco Bay, about three miles off the coast. We live in the only house on the island, which, of course, is edged with rocks.

At every high tide, wave after wave explodes against those rocks. A mighty storm can crack them together like bowling pins; even a weak storm can dislodge them. Amid the battering, however, many creatures thrive. They have adapted to the tumult here as others have adapted to the salinity of a marsh or the shifting sands of a tranquil beach.

Just as the marshy and sandy shores have specific places for certain forms of life, so does the rocky shore. Between the lowest low-tide mark and the highest high-tide mark, a rocky shore divides into six distinct neighborhoods. When the tide is out, a walk of about ten yards covers all of them.

Like a handful of jewels left by the falling tide, surf-polished rocks are among the treasures awaiting tidepoolers along rocky coasts.

I watch my step entering the first, a jumble of black rocks. A few feet from dry land, the rocks are rarely touched by the sea; they get wet only from spray at high tide. But the rocks here are slippery much of the time because of the kind of life they harbor: blue-green algae, among the oldest and most simple of organisms.

The black coating on the rocks is a slick, gelatinous substance that protects the algae by keeping them damp. Our visitors have skidded here so many times that in the kitchen we have put a sign that warns them on their way out: *Black Rocks Slippery!*

Beyond the black rocks is a rocky neighborhood regularly pounded by the waves but under water only at extremely high tide. Its regular inhabitants are periwinkles, which wander from their home grounds to the black rocks. There they feed on the algae by scraping them with a radula, a long, hard organ studded with hundreds of microscopic teeth. Periwinkles can stay atop a rock a month or more without water and can drown if they are forced to stay in water beyond their tolerance.

The periwinkles on these Maine rocks may have descended from kin that in Viking days drifted on logs from Europe to North America. Ever since, they have been heading down the coast at periwinkle speed. Reported in Newfoundland in 1861, they seem to have found a new port every decade or so. It took them until 1959 to reach Ocean City, Maryland.

As I leave the periwinkles, I start getting my feet wet in the barnacles' territory. The tide is going out, giving the clusters of grayish-white barnacles the first of their two daily exposures to air. To get food, they depend on the movement of the tides, having no way of moving themselves. They begin adult life stuck to a rock and that is where they stay. When the tide comes in, a barnacle will thrust out six pairs of legs that look

Gray periwinkle snails
hold onto the rocks using
adhesive mucus, while the
Forbes' common sea star
uses suction cup feet.

snails, and worms meander around the rockweed tendrils, which provide damp shelter amid the rocks dried by the sun.

Not all the residents of the rocky shore are so neatly confined to their neighborhoods. Limpets, for example, commute from one area to the next, searching for rocks coated with algae. Uncoiling a tooth-covered radula similar to a periwinkle's, a limpet hungrily scrapes away, sometimes so vigorously that it swallows bits of rock. Limpets stay attached to rocks even in a storm; when waves wash over a limpet's shell, the force just makes the creature tighten its grip on the rock even more.

The water is up to my ankles when I enter the next niche, characterized by Irish moss. Once in a while, men with long wooden-tooth rakes appear and begin wading along the rocky shore. They are mossers, followers of an old Maine trade, the harvesting of the reddish-brown seaweed known as Irish moss. They sell the moss for its carrageenan, a gelatinous substance used to keep ice cream smooth. I once came across a recipe for a pudding called blancmange, a dish that Chaucer mentions. It called for boiled-down Irish moss, cornmeal, and milk. I didn't try it.

The Irish moss grows thick upon the rocks, its curly fronds forming a shimmering under-

Where wave action is strong, blue mussels usually win the competition for space (above), crowding out barnacles and devouring algae. Growing in a dense mat next to the water's edge, Irish moss (below right) shelters many small creatures such as the moon snail.

like tiny feathers. Three pairs of legs will wiggle to pick up nutrients in the water while the other three will clean off the food and kick it into the barnacle's mouth.

A boat bottom is a safer place for barnacles, at least until the boat owner decides to scrape them off. Attached to a boat, a barnacle is out of reach of the sea stars and snails that prowl the shallows in search of food.

The barnacle carpet ends where masses of rockweed pick up. In the rockweed zone, long, rubbery strands of rockweed are strung with bladders which resemble tiny balloons. Children sometimes crush them to hear the popping sound, but without the little bladders, rockweed sinks and dies. The bladders buoy the rockweed when the tide is high, keeping the fronds upright and waving in the currents. At low tide, the tendrils of rockweed lie limp and tangled.

Standing or drooping, rockweed serves a large community. At high tide, prey and predator flit among the fronds, hunting and fleeing in the rockweed's shadows. At low tide, crabs,

water shrub exposed to the air only during the month's lowest tides. Intermingled with it are clumps of sea lettuce, whose broad leaves are bright green. Sheltered here is a myriad of unseen life. Tiny worms and crabs hunt for invisible food. Microscopic bryozoa, animals that cluster in colonies that look like delicate lace, array the stems of the seaweeds.

At the seaward edge of the Irish moss zone I am standing in knee-deep water. This is the last neighborhood that a wader can enter. It is sometimes called the Laminarian zone, after a genus of brown seaweed whose domain is the sea rather than the shore. The zone continues to the deep. Wading, swimming, or rowing in these clear waters, though, we can see animals that cannot live in the air: sea urchins, sea stars, jellyfish, fish, and, here in Maine, lobsters.

Bobbing in the waters around the island are the brightly colored buoys marking strings of lobster traps. The colors, like cattle brands, identify the owners. The lobster boat that hauls up the traps shows the same colors. And it better. Lobster-trap theft in Maine is taken about as lightly as cattle rustling in the West.

When the tide pulls the sea away from a rocky shore, some of the water remains in basins and crannies in the rocks. A short stroll along the Maine shore at low tide will reveal many such pools, each one subtly different from

Almost velvety to the touch, red or yellow blood stars may be found at low tide among the rockweed. A blood star broods its young in a pouch made by curling its arms under its mouth.

*Sea stars (right) come in
wondrous varieties, such
as the pentagon-shaped
horse star at left. The
northern sea star at right
has a whitish, disklike
sieve plate that takes in
water for the hydraulic
system sea stars use to
cling and creep.*

its neighbors. Some of the coast's most spectacular sea stars—purple and red sun stars and blood stars—prowl these pools to find prey.

Lining most tidal pools are mussels packed row on row, tighter than football fans in a crowded stadium. Infant mussels have a foot jutting from the newly formed shell. The foot, used to probe for a good anchorage, also helps to form the silky adhesive threads the mussel excretes. These threads stick to a rock and harden, connecting the animal to its permanent home. By opening its hinged shell and letting the nutritious tide flow through, a mussel eats.

Many of the eaters are eaten, however. Crouched by a tidal pool full of mussels, I have often seen a sea star making the rounds, searching for its next meal. To eat a mussel, a sea star first wraps its five arms around the black, shiny shell. The mussel clamps its shell shut as soon as it detects the predator, but it is usually too late. On the underside of the sea star's arms are hundreds of flexible tubes (called feet and easily seen with a small magnifying glass). Suckers on the ends of the feet tighten the sea star's grip. Exerting a steady pull, the sea star forces the mussel shell open slightly. Into the narrow slit goes the mobile stomach of the sea star, which engulfs the mussel's body and digests it.

Low tide or high, the rocks are patrolled by the great black-backed and herring gulls. We have come here early enough in the summer to see eggs in the nests that the gulls build on the rocks, and we have come late enough to see the adults teaching their fledglings to fly. Our island log is full of notes about birds: *dive-bombed by gulls along well path, where there are nests and chicks . . . Watched a chick emerging from shell . . . Spotted a new osprey nest.*

The guillemots are our special birds—small, sleek, and black with white wing patches. They have bright red feet and an air of elegance. Unlike the many screeching gulls, the little

When old enough to leave its cliffside nest, a guillemot hides among the crevices until its plumage has grown and it can fly. From among hundreds of clamoring chicks, guillemot parents have little trouble picking out their own offspring.

their wings, splashing and moving in circles. The intricate courting ritual continued until the middle guillemot suddenly flew to a narrow ledge on the cliff. The second bird followed. They mated. As the domestic drama ended, the third bird flew to another part of the cove.

On another day, we saw two birds taking turns flying up to the cliff, landing on the lip of a crack and vanishing inside. Usually the birds carried something in their bills.

I couldn't suppress my curiosity. I climbed up, lay prone on a nearby outcrop, and, when both birds were gone, peered into the darkness. In a rock cleft, atop a scant bed of grass, nestled a downy guillemot chick.

A nearby island is the home of hundreds of cormorants. The skyline of the cormorants' island reminds me of the battlement of a medieval castle: a row of black birds, wings spread and bent, as if they had elbows. The birds stand so close that wing almost touches wing. At our approach, some fly off to their fishing grounds. They swim about, their necks snakelike, bills pointed upward, looking as graceful in the sea as their brethren look awkward on the rocks, drying their wings in the cormorant version of arms akimbo.

The wind is to our backs and we are grateful, for no place I have ever visited stinks more potently than this cormorant rookery. The stench is overpowering. Scattered about the flat top of the island are hundreds of meager nests, which seem to be made principally of guano. Very fresh guano.

The cormorants' name comes from the Latin for "sea ravens," and, like ravens, they can croak loud and long, especially when we approach their nests. We do not stay long. For humanitarian reasons, we don't want to frighten the nestlings that are lurching about, naked

guillemots are silent and easy to watch one at a time. There are never more than eight or ten swimming in a little cove near a sheer cliff that thrusts out of the sea. The wall of rock is scored with cracks, making it look as if it were made of a giant's building blocks.

The great chunks of granite change color —gray, beige, golden—as the sun's shifting light plays on the minerals captured in the stone. Up close, mica and feldspar glitter, especially along the fault lines where the rock has cleaved. In these cracks, a grudging gift of the rocky coast, the guillemots build their nests.

One day we watched three birds bob in the cove, some distance apart. One dived swiftly, disappeared, then popped up alongside a bird that seemed oblivious to the performance. Then another guillemot went through the same actions and appeared on the other side of the aloof bobber. The two flankers began rapidly beating black chicks that look like animated rubber toys. And for reasons that also have to do with our noses, we soon leave.

Another nearby island, with fewer rocks, seems to belong to terns. We rarely see terns on our island, and our gulls seem to avoid the terns' island. But cormorants often line our rocks with their wing-to-wing formations, and our gulls often fly out to the cormorant rookery and join in the fishing.

We have given up trying to sort out the ways of the terns and the cormorants and the herring gulls and great black-backs. Although individual gulls squabble continually, we have seen none of the widespread aggression that we have read about in scientific observations of gulls. Facing real or imagined danger, they all act together. Chicks waddle off to the nearest hiding place in the rocks. Adults soar up from the rocks in shrieking clouds of outrage. When a human being—or a low-flying hawk—nears a nest or a wandering chick, a common alarm goes off. All the gulls, regardless of species, cry out and fly down to harass the intruder.

Aloof in a huge bramble of a tree-top nest, the ospreys seem more patient with their young than the gulls are with their young. We have yet to see a new osprey fly from its nest. But we have seen, again and again, a young gull flap its wings, stagger along a flat rock, glide, rise—and crash into the sea.

Sometimes an adult, flying nearby, will dip down, land near the young bird, and apparently escort it back to the rocks. Sometimes a crash-landing bird gets no notice, except from us. We wonder why one bird gets attention and another does not, and then we begin watching a different bird about to take wing. Questions and answers fade away in a place where we can intrude and behold and discover but never enter. 🐚

Small and ducklike, black guillemots feed on fish and slippery rock eels. One observer likened them to large black-and-white bumblebees because their wings move so fast in flight.

GULF

Brown pelican in Florida mangrove

I have concluded from my travels that the U.S. Gulf Coast is composed of the following: gulls, alligators, blue-green water, oil wells, goodoleboys in gimme caps, condos, RVs, shrimpboats, pelicans, herons, pastel motels, mosquitoes, girls named Darlene, blue-green water, hurricane stilts, egrets, catfish, raccoons, middle-aged women in pink shorts, humidity and blue-green water.

That blue-green water meets the land in an arc 1,632 miles long. It is America's full-of-surprises Gulf Coast—seldom seen salt marshes and mangrove shores not far from beautiful sandy beaches that draw vacationers, fishermen, boaters, developers and beachcombers. This coast around the Gulf of Mexico harbors immense riches—everything from oil and sulphur to stupendous colonies of shrimp and crawfish. Some of the best of its natural beauty is along wind-sculpted barrier islands.

On a hot morning in early fall, Paul Eubank guided his Park Service pickup onto the sand a few yards from the surf and headed south. A ranger at Padre Island National Seashore since 1974, Eubank is a somewhat taciturn Texan who is content where he is. "I like the beachcombing and the clean salt air and the

If seashores had a symbol, it might be the ever-present birds along the water's edge, such as these black skimmers at a nesting colony (left). In search of fish, adult skimmers fly just inches above water (right). When a skimmer's bill touches prey, it snaps shut.

Sea oats are durable and vulnerable at the same time. Their long roots run vertically and horizontally, knitting shifting sand dunes together. When people walk on the dunes, however, they can easily rip out the roots, and the whole knitting process must begin afresh.

openness here, the sense of freedom," he said. "And I like being the Lone Ranger. You know what they say, a bad day at the beach is better than a good day someplace else."

Padre is the longest (113 miles) barrier island in the United States. The 80 miles of it that are run by the National Park Service range in width from 1,000 yards to three miles, numbers that are subject to frequent amendment by wind and storms, sand deposits and erosion. Extending from Corpus Christi south to the Mexican border, Padre fronts the gulf on the east side and a shallow bay called Laguna Madre—Madre is never far from Padre—on the west.

The island squeezes a variety of wildlife into what is essentially a long ribbon of a sandbar fancied up with dunes, rainwater ponds, and grass flats. The beach on the gulf side is a raucous neighborhood of gulls and terns and ghost crabs and sanderlings. On the lagoon side it's fiddler crabs, herons, white pelicans and more gulls. The grassland between them is fine habitat for coyotes, pocket gophers, diamondback rattlesnakes, and, in winter, sandhill cranes and peregrine falcons. Also gulls.

Bobcats have been seen occasionally and, strangest of all, so have enormous blue-gray antelope called nilgai. Eubank has spotted the elk-sized antelopes a half-dozen times and described them as "skittish but huge; looked to me like the big ones weighed 800 pounds." The short-horned beasts apparently swam or waded across the lagoon after ranchers brought them to Texas from India as a game animal.

Eubank drove slowly down the beach boulevard, which was devoid of people except for a few fishermen trying for redfish. Sanderlings quickstepped just behind the receding waves, pausing every few seconds to stab a bill into the sand. Terns in platoon-sized detachments worked their way down the beach. Black

skimmers, most aptly named of birds, glided inches above the water with their scissorlike red beaks open, taking potluck in the surf.

Skimmers are the lone members of the bird tribe equipped with a lower mandible, or bill, longer than its upper counterpart. This permits them to skim gracefully at wavetop level with the lower mandible in the water, scooping up small fish and crustaceans and then snapping their head down when the bill clamps shut. In between feeding forays, they congregate in clusters on sandbars, heads cocked into the wind. When tired, the white-breasted skimmers sometimes rest by flopping down on the beach with their bills flattened out, as if they could skim no more.

Skimmers nest on islands in Laguna Madre and, more perilously, in a fenced-off area along the causeway from Corpus Christi to Padre. Donna Shaver, a resources management specialist on the park staff, said that people often park their cars so close to the fence that it dis-

turbs the nesting skimmers. The mothers leave the nests when people come too close, and as a result the eggs often overheat.

The birds we love to hate, the chronically pushy gulls, were gathered along the beach in boisterous crowds. At Padre the relatively small laughing gull (actually more of a chuckler) is joined in fall and winter by larger herring and ringbilled gulls. Gulls make a good living on crustaceans and fish and insects, supplementing their menu with selected *hors de garbage*. Like coyotes and alligators, gulls ask few questions about food except how much and how soon. They share another trait with coyotes and gators: they need a better press agent.

Gulls take an unseemly amount of abuse for doing what comes naturally. Children throw stones at them; adults, upset when gulls sully their boats and cars, sometimes poison them.

Padre Island ribbons along the Texas coast for 113 miles, sheltering the mainland and a host of wildlife. Amid the sea oats, a clump of yellow camphorweed thrives.

Gulls make sorties along the sea's edge for crabs (right) and other crustaceans. Soon this young laughing gull (above) will begin its own forays for food.

Wildlife managers get annoyed because gulls push terns from their nest colonies. Gull numbers have exploded everywhere, making their bad habits—such as gumming up airplane engines—more conspicuous.

Only recently have gulls acquired a cadre of scientific defenders who find them fascinating. At their nesting colonies, they stake out modest territories that the males defend ferociously. Gull couples mate for life and normally

display a commendable faithfulness; though they often separate in winter they find each other once again, by some means that confounds biologists, the next spring. They communicate through a half-dozen calls with specific meanings, and apparently make judgments about each other's trustworthiness—alarm calls emitted by some gulls are ignored by others in a colony. A California researcher discovered recently that gulls recognized him as an individual; by donning masks and various outfits he proved that they responded—negatively, as it happened—to his particular facial features.

In the air gulls opt for individuality. The members of a milling flock appear distracted and disorganized, which is how several dozen Padre Island laughing gulls were deployed as the Lone Ranger's pickup passed. A few yards farther on, Eubank stopped and got out to inspect a mother lode of seaborne trash.

Experienced beachcomber that he is, Eubank ignored the dozens of colorful plastic containers and zeroed in on the higher-caliber stuff: two damp coconuts, direct from Central America or Mexico, and a 19-inch Sylvania television set. "I guess the best thing I've found was an inflatable boat with a 40-horse engine," he said, "but I've found handguns, a diamond ring, pieces of a helicopter that went down and, once, a body. We think it was a fisherman who drowned off El Salvador."

A lightbulb lay half-buried in the sand next to an egg carton. Nearby was a tree trunk, perhaps five feet thick and 25 feet long. "From Yucatan, maybe," he said. The trash's arrival on Padre's beach is the consequence of winds and currents that converge there, especially in the springtime and fall, and wash much of the gulf's accumulated garbage ashore. By contrast, the beaches across the gulf on Florida's west coast are by-and-large free of such wave-borne debris.

"The currents here are always inshore," said Eubank. "You could put a note in a bottle and throw it in the gulf and nobody but you would ever see it. It would keep floating right back."

Eubank walked a few yards shoreward and squatted by a two-inch-wide, arrowhead-shaped ghost crab burrow. Dozens of others pocked the beach nearby. Dubbed ghosts for their pale color, the crabs feed at night between the tide line and dunes. Eubank pointed to the corpse of a tern next to one hole. "The crab dragged it here to eat. Must have been quite a load."

Ghost crabs seem able to coexist with trash, but it is often fatal to other shore critters. Brown pelicans become entangled in drifting plastic lines when they dive for fish, and sea turtles sometimes swallow plastic bags. For humans the most dangerous flotsam is metal

line. Over time seven miles of its east end eroded and three new miles were added to its west end. Now the whole island lies in the state of Mississippi.

At Padre, the gentle surf deposits sand and enlarges the beach, but storms and bigger waves tear at the beach and blow its sand to the inshore side, where it is slowly filling in the lagoon. The wind-built dunes persist until a storm pushes them across the island. (The average pace is 35 feet a year.)

"Hold on," Eubank said as the pickup reached a place called Devil's Elbow. "For the next 10 miles I'm going to bounce your kidneys." The composition of the beach changed here, from sand and small shells to something that approached a shell-collector's paradise.

Eubank pointed out handsome varieties—neatly symmetrical sand dollars, coquina clams with their decorative bands, the spikelike shells known as Sallé's auger, fan-shaped scallops, and the curling shell called baby's ear. Beach-walking here was like treading a railroad bed. The large shells come ashore at this particular stretch of beach because the currents flowing parallel to the shore converge here. The strand formally known as Big Shell Beach has been called the shell capital of Texas.

Bragging rights to the Gulf Coast's "best" shells, of course, have to be shared with places like Florida's Sanibel and Captiva islands, where people go out at night wearing miners' lamps on their heads to sift through seashells, and local grocery bags are printed with shelling guides. About 400 species of shells wash ashore at Sanibel, the most bountiful treasures arriving after particularly bad storms.

Like much of the Gulf Coast, Sanibel was first explored by the Spanish. Ponce de Leon headed into the area in 1521 looking for a magical fountain that kept one "forever young."

For a twist on the traditional outing to a sunny beach, try strolling the sand after dark. Then the world belongs to raccoons and coyotes creeping across the dunes, sandhoppers and ghost crabs (right) scampering in the shadowy moonlight.

drums that sometimes contain toxic chemicals. John Hunter, superintendent of the national seashore, said that 285 drums rolled onto the Padre beach in 1988.

Since the park lacks the money and manpower to keep the beach clean, conservation groups have organized cleanup campaigns in recent years. In April, 1988, during a one-day "trash-off," 2,106 plastic items, 406 glass objects, and 369 pieces of metal were gathered at Padre. Needles averaged 11 per mile. Debris came from countries ranging from China to Sweden to Saudi Arabia.

The same currents that deposit such garbage also bring tons of sand. Like many barrier islands, Padre originally formed when a series of offshore sandbars were connected by sand carried by the currents that run parallel to the coast. Padre is perpetually being sculpted by forces that simultaneously build and destroy, the same forces at work on other barrier islands. A sandy Gulf Coast island named Petit Bois, for example, used to straddle the Mississippi-Alabama state

Another group of Spanish explorers was discovering Padre in 1519, calling it Isla Blanca (White Island) for the color of its sand. In 1800, the Portuguese priest who gave his title to the island brought a herd of cattle there. Ranching remained the main business on Padre until the 1900s, though builders made several passes at it as a potential resort. The last attempt to develop it, in the 1950s, ended with resort communities on either end of the island and the establishment of the national seashore in between.

The air of mystery and romance that often suffuses islands is part of Padre's lore as well. One legend (also heard at Nag's Head on the Atlantic Coast) says that pirates once lured ships aground by walking donkeys with lanterns on their backs along the shore at night. Seamen supposedly thought the gently bobbing lights were the masts of other vessels. Smugglers have long been partial to Padre as a convenient entry point; today's contraband is most often drugs and illegal aliens.

Despite its evolution from pirate coast to federal preserve, Padre has remained defiantly, incorrigibly Texan. Texans are accustomed to driving their pickups on beaches, so driving (in four-wheel drive vehicles) is permitted on all but a small section of the beach at Padre. So is tent and RV camping, up to the dune line, with the result that on summer weekends the beach often looks like a linear campground.

None of this was accidental. "This is the spiritual headquarters of free enterprise," said Ed Harte, a Corpus Christi publisher who campaigned for the park. The Texas legislature, which had to give its blessing, made sure that oil company mineral rights were respected; oil wells still persist on the island. "It's the Texas ethic," Harte said. Likewise beach driving. "If they ever tried to change that, there would be a hysterical uproar," he said.

The Lone Ranger turned onto a trail that climbed over a 25-foot dune to a grassy flat. A northern harrier circled overhead at trolling speed, overlooking a jackrabbit crouched in a tumbleweedlike indigo plant. Eubank stopped beside the decaying remains of a rancher's old corral, but instead of horses or even nilgai the corral was occupied now by a row of great blue herons perched on fenceposts, as alert and motionless as sentries at attention.

The strip of beach ended in a pair of stone jetties on either side of Mansfield Channel, the 200-yard-wide waterway at the park's southern border. "We're only 30 miles from Mexico here," Eubank said. "Illegal aliens sometimes swim across and walk up our beach."

The jetty protruded so far into the gulf that a lone fisherman near its end was a barely visible speck. We walked midway down the breakwater, where there was a chance of spotting one of the Gulf Coast's most celebrated creatures—a

A beachcomber's bounty washes ashore from the sunny tip of Texas to Florida's Captiva and, left, Sanibel islands. Discovering a rare lion's paw with both valves intact (below) would elate any searcher.

Growing up to two feet long, the Florida horse conch is easily the largest snail found in American waters. As the mollusk inside grows, it steadily deposits new crystals of calcium and makes a bigger shell. Each stage of shell growth is marked by a knobby ridge.

sea turtle. Several types of turtle once flourished here, from the 100-pound Kemp's ridley to the 1,200-pound leatherback, described by a man who encountered one as "Volkswagen-sized." The midsized models are the green (250-400 pounds), loggerhead (200-400), and hawksbill (80-140). For centuries, sea turtles have been killed for their meat and shells, their sandy nesting grounds have been ruined, and their eggs stolen. Sea turtles also drown in shrimpers' nets and strangle on trash bags.

The predictable result of this exploitation is that all five species are now classified as either threatened or endangered.

Biologists believe the females return to the beach where they were born to lay their eggs. If the beach has been developed in the meantime, however, the result can be fatal for the hatchlings. When they break out they instinctively crawl toward the lighter part of the sky, normally over open water. If artificial lights are present they crawl toward them instead—and perish.

Eubank eyed the surf for 30 minutes before his patience was rewarded. First one and then another sinuous turtle head popped out of the water like a periscope. They were green turtles, feeding on the sea grass that grows in the shallows near shore.

Biologists are in the dark on precisely how sea turtles find their way back to their native beach, if indeed they do (perhaps by imprinting the chemistry of the sand). Researchers at Padre, led by Donna Shaver, are trying to unravel some of the answers while attempting to save the Kemp's ridley by establishing a new breeding population of the turtles at Padre.

Every year from 1978 to 1988, eggs laid by ridleys in Mexico were placed in boxes containing Padre Island sand and incubated. The hatchlings were released on the beach at Padre and then recaptured in the surf. The idea was to imprint Padre sand on the infants. After recapture they were raised until they were a year old and roughly plate-sized, then were tagged and released for good off Padre in the hope that they would return there to nest. Through 1989 no turtles had returned to Padre.

The best time to look for shells (lower left) is at low tide, especially after a storm. The Gulf Coast shores are dotted with sundial shells (left), a variety so perfectly symmetrical that it is also called the architect shell.

Nevertheless, Donna Shaver and company are like prospective parents awaiting nature's verdict. They faithfully patrol the beach four times weekly during the April-to-August nesting and incubation season in search of eggs and flipper tracks. "I enjoy working with turtles because there is still so much mystery about them," Shaver said, "and because it's us—people—who caused their decline. I want to do what I can to bring them back."

The slender grassland that fills the island's midsection behind the dunes is surprisingly rich in wildlife. Thirteen kinds of snakes, including two that command deference—diamondback rattlers and massasaugas—live in the plant-carpeted interior dunes. The pocket gophers the serpents prey on construct intricate burrows and sensibly do their dining underground, nibbling roots that penetrate their passageways.

As the wind moves dunes across the flats, grass-fringed troughs are left behind, small depressions that fill with rainwater and become ponds where ducks assemble. Flocks of greater sandhill cranes settle in the grass beside the ponds after the first north-country cold snap. Tall, clamorous and unmistakable with their gray plumage and red foreheads, the cranes seem to do everything full-throttle, from their foot-flapping water dance to the clarinet-like calls that can be heard from one side of Padre Island to the other.

The peregrine falcons that also show up in winter are, by contrast, low-profile, as silent and stealthy as the cranes are raucous and high-spirited. Padre's peregrines are under study by scientists tracking their migratory routes. The park staff keeps mum about peregrines to foil humans who might bother the birds.

The sand that blows across the interior dunes also accounts for the disappearance of live oak forests that once dotted Padre. The sand smothered the oaks' acorns, perhaps gradually

or perhaps in one powerful storm. All that remains is a few trees so scrawny and beleaguered that they look as if a hummingbird landing on a branch would topple them. Their roots bared by the wind, the tired trees lean toward the mainland as though trying to escape. Rangers call this frail relic "the Padre Island National Forest." Live oak forests have fared better at Gulf Islands National Seashore, where the sturdy trees flourish in a sandy ridge.

The plants that survive on Padre's dunes and the flats behind them have evolved strategies to offset the constant onslaught of wind and sand. The tall sea oats that wave like willows on the dune line have the handy ability to continue growing when sand piles up around them, a circumstance fatal to many other plants. They take hold with such tenacity in sand that they are used as dune stabilizers. The roots of the delicate looking purple-and-white beach morning glories continue the process.

Whitestem wild indigos, which provide shelter and nest sites for several grassland animals in spring and summer, turn dusty gray in fall and then break off at the stem. The indigos ensure their survival by dispersing their seeds as they roll, tumbleweedlike, across the island in fall and winter.

The Laguna Madre, which separates Padre from the mainland, is an ecologically rich estuary that serves as a nursery for dozens of species of fish and invertebrates, including tasty brown shrimp. The thick clumps of shoalgrass on the bottom of the shallow bay provide both spawning grounds for fish and shrimp and winter feed for ducks, especially redheads. Some of the bay-dwellers are runt-sized due to the high salinity caused by the absence of rivers flowing into the estuary and by evaporation.

The more open bays on the gulf between Louisiana and Florida have a much lower salt content than the Laguna Madre. In dry years, Laguna Madre becomes three times as salty as the gulf, and major fish-kills result there.

The white pelicans that work the edge of Laguna Madre are year-round residents who nest in America's only saltwater white pelican colony, on South Bird Island. Unlike their showier brown cousins with their penchant for diving hell-bent into the surf for a meal, the whites favor a communal feeding technique. They form up in flotillas and herd fish into dead-end shallows, like Texas cowhands herding dogies, before smoothly scooping their prey into their bills.

Pelicans have never quite mastered walking, and their takeoffs are so awkward that every one looks like a maiden voyage; but in the air

Scanning the grassy interior side of Padre Island, visitors might see a stealthy bobcat (left), coyotes, raccoons, even an exotic antelope-type creature called a nilgai, which nearby ranchers imported years ago.

White pelicans form a communal feeding circle to herd fish into shallow water so they can easily scoop them up. The pelicans live year-round at Laguna Madre, a salty body of water that separates Padre Island from the mainland.

the clumsiness vanishes in a flight profile so smooth that it approaches avian elegance. With their nine-foot wingspans the whites are slightly larger than the browns, and they have also adapted better to modern life—the DDT that nearly wiped out the brown pelican population had but a slight effect on whites.

Biologists fear for the Laguna Madre colony's survival because of the pelicans' well-documented touchiness about even the most minimal intrusions. If people show up on their

home island during nesting the whole band could well decamp, and the number of alternative sites is fast dwindling. One colony departed *en masse* some years ago, apparently because a sign was erected on its island. The sign identified the island as a bird sanctuary.

On my last afternoon at Padre the word came down early: "Major storm in the gulf. Condition yellow." A yellow alert, first in the three-stage hurricane warning sequence, signalled rangers to close the park and get the people out. The storm was 500 miles offshore and moving toward the Texas-Louisiana coast.

Nobody on the Gulf Coast takes hurricanes lightly. The 1900 hurricane at Galveston, a city built on another barrier island 200 miles up the coast, was the most catastrophic natural event in American history with about 5,000 deaths. Hurricanes in Texas are no rarity. In 1961, Carla gouged 10-foot-high cliffs out of the Padre dunes. Allen in 1980 cut 26 inlets in Padre.

Four cars and a motorcycle remained in the visitor center parking lot when ranger Darlene Carnes posted a notice at 2 p.m.: "The Weather Service predicts the storm will be at hurricane strength by 5 p.m. It is expected that the highest swells will pass over the dune line."

"Don't you love it, the excitement?" Carnes said before going off to find the missing visitors. I had to admit that I did.

The storm now had a name—Jerry—and 80 m.p.h. winds. Though it wasn't expected to strike the coast until the next day, the rising tide might flood the causeway to Corpus Christi before then. (I received permission to stay through the afternoon.)

Superintendent John Hunter arrived and together we glared at the surf. The gulls and sanderlings on duty at the tide line seemed unbothered. The waves were higher than normal, but only slightly. Visitors trickled back from the beach one by one and drove off, retrieving their $3 entrance fees at the gate. At 4 p.m. only one car remained.

But Hurricane Jerry, alas and thank God, caused no more excitement at Padre Island. When I crossed the causeway at 5 p.m. the water was still two feet short of it. That night I got up frequently to peer out the window of my beachfront motel room, but nothing much was going on. Jerry struck Galveston the next day. Three people riding in a pickup on a gulfside road were swept away and killed.

When I left the following morning the gulf had returned to its customary blue-green tranquillity. The skimmers and gulls were working the surf. The herons were on their fencepost perches, fleets of white pelicans were floating in Laguna Madre, and the ghost crabs were scudding furtively from hole to hole. Paul Eubank was right. Come trash or high water, this slice of the Gulf Coast is still a better place to be than most anyplace else.

Fierce 90 m.p.h. winds redeposited tons of sand when Hurricane Allen's fury hit Texas in 1980. Allen created the highest tides in 61 years, but many residents of the Gulf Coast seem inured to such superlatives— they withstand several hurricanes in a lifetime.

avid Richard halted the airboat and spread his arms as if to embrace the scene before him. The October morning was still except for the barely audible splash of a fish breaking the smooth surface. A checkered quilt of marsh grass, varying in color from light brown to rich emerald green and interspersed with dark patches of shallow water, unfolded toward the gulf to the south. "This is my marsh postcard," Richard said. "They could bury me right here and that'd be fine." A dark-bodied, red-billed gallinule suddenly shattered the silence by sprinting across the surface of the pond like a commuter racing for a train.

Richard (pronounced *Ree-SHARD* in Cajun country) works as a biologist at Rockefeller Wildlife Refuge, an ecologically affluent, flat-as-a-spatula marshland preserve that extends 26 miles along the Gulf Coast in southwest Louisiana. Such marshes fringe

Salt marsh country plays a crucial role for birds such as egrets and ibises (right). The attraction? Food in abundance. An ibis pokes its bill into the muddy water to adeptly nab crawfish (left).

virtually all of southern Louisiana, though they are less common along the rest of the Gulf Coast. They are a powerful presence—humid, peaceful, abundantly productive acres of marsh grasses swaying in the gentle gulf winds.

Humans need a boat to navigate the marsh, and the hardy birdwatchers and sportsmen who make the journey usually spend half their time slapping away the ever-present mosquitoes. Yet the trip is worth it. It's the sense of careless abundance that gives the marsh its sensory impact, a richness of life that only gradually reveals itself. Marshes are deceptive, at times eerily quiet and at other times aclatter with the screech and bustle of birds by the tens of thousands.

Richard guided the boat through the marsh grass, arriving almost without warning at the open sea. Overhead there was a band of laughing gulls milling about in disorderly gull-like fashion; beside us was a narrow shell beach few people ever see, because of the marshy terrain that adjoins it. The water was blue-brown.

When the male fiddler crab brandishes his large claw and a female accepts the invitation, the two enter the male's burrow to mate. Fiddlers got their name from the male's large fiddle-shaped claw.

"No Florida water here," he said, "too much organic matter in it."

A dozen shrimpboats spread their outrigger nets a mile offshore. Their quarry was the brown shrimp that spawn in deep water (to 300 feet), grow bigger in the cozy shelter of the salt marsh (they almost double in size weekly on the rich soup of marsh waters) and then swim offshore to lay their eggs. Found from New Jersey to Brazil and all along the Gulf Coast, the succulent brown shrimp, together with their pink and white kinfolk, are the mainstays of the U.S. shrimp industry.

Richard bounced the boat off the beach and coaxed it through a half-mile-wide band of salt marsh creased by meandering bayous. He careened over thickets of grass, admiring great fleets of long-necked avocets.

Despite the airboat's clangorous motor, it is the easiest way to travel around and see such sights. Richard roared across a pond and calmly kept going when he ran out of visible water, bouncing over mats of floating vegetation and weaving through the thick grass like a Manhattan cab driver dodging traffic.

Amid a bright green and gently undulating carpet, Richard stopped the boat. He got out and took a few cautious steps across a mat of vegetation so thick that it was impossible to see the foot-deep water on which it floated. Sinking a few inches with each stride and extending his arms for balance, he looked like he was walking on a water bed, which in a sense he was. "To me this is solid," said Richard. "You learn where to step." He has walked out of the marsh several times when boats broke down. "Mostly it was a walk, but I swam some too. You can definitely float your cap in here if you're not careful."

Scores of species thrive in the dark water, the fiddler crab being among the best adapted of them. It can survive exposure to salty seawater and, under the edge of its shell, has a primitive

lung that helps it breathe. Fiddler crabs can go without air for long periods, however, as they wait out high tides in their burrows.

To deal with the marsh's mud, the fiddler uses special bristles on its mouth parts to sort out pieces of food. It swallows the fine particles and collects inedible ones in a special chamber, eventually spitting them back into the marsh.

The black, loamy material—more roots and plant matter than soil—is the key to the marsh's richness. Richard plunged an arm into the wavy mat and brought up a handful of it. "It's sometimes tough to figure out what's dirt and what isn't," he said. The thick tangle of decayed vegetation feeds the tiny organisms, which in turn support the insects and aquatic creatures and so on up the chain to ospreys and herons.

A marsh develops when the stems of marsh grass slow the currents enough for sediment to gather around their bases. In Louisiana, the currents are usually from rivers flowing south to the gulf. The marsh expands as the grass traps ever more sediment.

"This is the most productive piece of landscape I can imagine," said refuge biologist Larry McNease, who has lived and worked here 22

A skittering mass of fiddler crabs emerges from the marsh grass at Appalachicola Bay in Florida. When the tide comes in, fiddler crabs take refuge in their burrows and plug the entrances with mud.

years. "They talk about the Everglades, but that seems almost sterile next to this."

While the salt marsh ribbons directly along the gulf, the adjoining marsh on the inland side is more brackish. Pure sea water holds at least 36 parts salt per thousand parts water; brackish water holds less, down to just above one part of salt. The degree of dissolved salt helps determine the kind of life forms that thrive here —fiddler crabs, muskrats or gators; wiregrass, cord grass or three-square grass, also known as swordgrass. The two-to-three-foot wiregrass has a somewhat swirly look because its leaf edges twist back on themselves and curl together. As wiregrass dies and decomposes, it becomes a vital part of the food chain.

Swamp rabbits, opossums and coyotes tread levees next to canals originally dredged by oilmen; otters, muskrats, and raccoons all find a niche in the brackish part of the marsh.

As Richard propelled his boat through the marsh, a cloud of bluewinged teal and a flock of graceful white ibises took wing. The biologist's eyes flicked systematically around a horizon aswarm with birds. He issued bird identifications with machine-gun velocity: "Least bittern —see the flash of rusty orange? They're shy, you don't see them much. Glossy ibis—see the shine? Black-necked stilts. Oh, and there's a flock of red-winged blackbirds. We have them by the absolute millions."

Its strategic location between two major bird migration routes makes the marsh the center of a year-round avian extravaganza. Great flocks of ducks and geese blacken the sky in November and linger through the winter. Stilts, ibises, mottled ducks, herons, gallinules and many others nest at Rockefeller and other Gulf Coast refuges in spring.

And the marsh is thick with songbirds. The marsh wren, for instance, uses leaves of the ubiquitous marsh grass to weave its ball-like nest. First the male has a try at it, building three nests or thirty. When the female arrives, she spurns these efforts and starts from scratch, spending five or six days constructing a softer nest. The male's "nests," however, don't go entirely to waste, since rice rats or square-backed crabs may use them for a resting place.

About the time the young birds hatch, the marsh seems to explode with other new life —insects and grasshoppers and greenhead flies buzzing in abundance. During a normal high tide, the birds flock to the marsh to nab such insects. But when a particularly high tide comes in mid-summer, the birds have even easier pickings. Then insects are forced to crawl higher on the marsh grass to escape rising waters, and the swaying stems become top-heavy with crawling creatures. Grateful wrens and other birds gorge themselves.

Many songbirds—sparrows and tanagers, vireos and grosbeaks and buntings—make rest stops in the scattered islands of trees in the marshes before soaring south in fall and after returning in spring. McNease has been trying for years to record the throaty bellow emitted by alligators in mating season. "But I can never get

Marsh grass makes a perfect spot to conceal a least bittern and its fuzzy youngsters (right). This furtive bird prefers slipping through the reeds instead of flying. It is so secretive that its soft call—coo, coo, coo—may be the only clue to its whereabouts.

The great white heron's fishing technique— standing motionless in shallow water until prey comes along—pays off when it snags a fish.

it clear on the tape," he complained, "because the birds always drown the gators out."

The songbirds that arrive in spring on the Gulf Coast have stored up their energy for the 600-mile flight from Central America by eating more food. Even this instinctive preparation doesn't always help when birds run into a storm bearing headwinds that force them to labor twice as long to finish their voyage. The result is a spectacle known to birdwatchers as "fallout."

Drenched and heavy-winged, the birds are so exhausted by the time they reach the coast that "they fall from the sky like raindrops," said McNease. "You see more birds in two days than you normally do in a month."

Farther inland, past the salt marsh and the brackish marsh, comes the freshwater marsh, the habitat of water hyacinth and alligators. Many marsh-dwellers, including alligators and birds, can make a living in more than one of the three life zones.

Among the most abundant creatures are the crawfish, also known as crayfish and crawdads. Catching and eating the crustaceans is a

full-scale business in Louisiana. The crawfish survives the changing water levels by digging a burrow up to three feet deep in the marsh's muck, then plugging the entrance and hiding inside until the water subsides.

Buzzing down a winding bayou, Richard pointed at a gray-brown head protruding from the water. "Nutria," he shouted. A nutria resembles a small beaver that has been fitted with a rat's tail and a set of dirty orange teeth. Equally at home on land and water, nutrias reproduce with abandon: an average of ten babies are born to each pair every year. Nutrias were originally brought to Louisiana from South America by Edward A. (Mr. Ned) McIlhenny, the Tabasco sauce baron who almost singlehandedly created Louisiana's network of wildlife refuges, including Rockefeller.

When it came to nutrias, however, Mr. Ned's normally fastidious plans went awry. The 150 captive nutrias that escaped from his compound in 1940 multiplied so recklessly that by the 1960s they numbered in the millions. While nutrias became a pest in parts of Louisiana, greedily chomping marsh plants along with

farm crops, they also were the main contributor to the state's recent $17 million fur harvest.

After rounding a bend in a muddy pond, Richard abruptly splashed to a halt and pointed. "See it, there on the bank?" he cried happily. "Gator, about a 10-footer. Just taking a little sun." The alligator density at Rockefeller refuge is the highest in North America, with a population estimated at 12,000 to 15,000. Most of the gators were under water that day—"down in their holes because it's getting cool," Richard said.

The refuge staff, led by manager Ted Joanen and Larry McNease, has been studying alligators for years. It was Joanen who showed that the sex of young gators is a result of the temperature during the 65-day period the eggs incubate: at 93° F or above, all the babies are male; seven degrees lower they're all female. Rockefeller biologists are forever measuring and studying gators, a pursuit that puts them squarely in the wake of Mr. Ned McIlhenny.

Mr. Ned was dismayed by the mythology that had long muddied alligator science— allegations that they were as fast afoot as a thoroughbred horse, for example, or that they lived 200 years. He cleared some of the water in

Farther inland, the salt marsh shifts to brackish marsh, and the rumbling roar of huge alligators can more often be heard.

Next page: A graceful flotilla of white pelicans moves along the channels of a Louisiana delta.

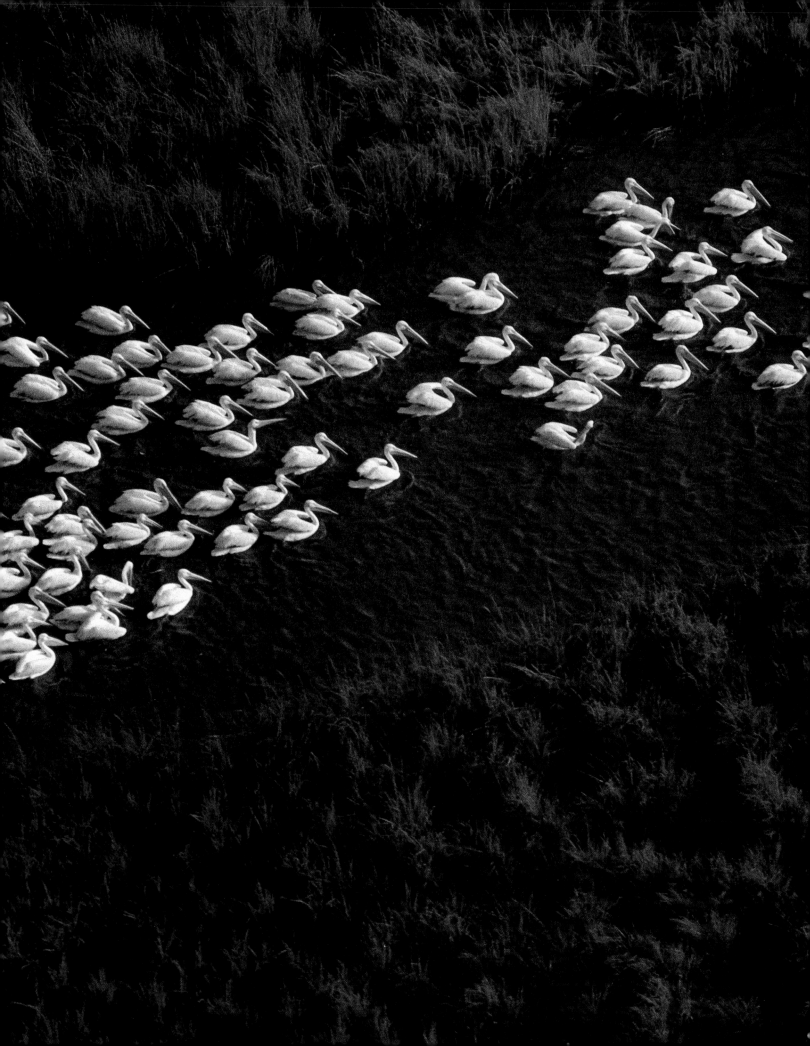

a book he produced in 1934 called *The Alligator's Life History*. McIlhenny described the flat-topped nest of dry plants a mother gator erects in the marsh (the similar house that a muskrat builds has a round roof) and reported that when an irritated mother once caught him too close to her nest he calmed her with soothing clucks.

Richard wove his way through a stand of cattails where a dozen egret heads and necks were discernible as S-shaped flashes of white amid the stalks. A squadron of red-winged blackbirds bustled busily by overhead while a marsh rabbit darted along a levee. The boat whined around one last corner and the handsome buildings of the refuge headquarters materialized. Nine white-tailed deer grazed on a meadow a hundred yards away.

Just before sundown I climbed an observation tower on the eastern rim of the refuge, the highest point for miles. The elevation was about 30 feet, but it felt like 80. From there it was easy to see that the mark of man permeates this marsh. A dozen oil company canals run straight as a column-rule north-south and east-west, intersecting at junctions that look like crossroads on the prairie. Beyond the canals the rough rectangles of light green, dark green, yellow-green and green-brown marsh grass shimmered in the afternoon light. The dark ponds flecked with the white glint of wings had the refreshingly imprecise configurations of nature.

I made a slow, 360-degree swing with my binoculars. Three alligators, their cool predatorial eyes riding the surface of the water, were parked in the canal, patient as old fishermen. An oil rig reared up to the west, and out in the gulf I could see the hazy forms of shrimpboats heading home like a fleet of ghost ships. The only sounds were natural, the reedy trill of wading birds and the soft splash of surfacing fish.

It struck me that things seemed in balance here, mud and grass and muskrats, shrimp and teal and gators. It was an efficient system: everything had a place, nothing was wasted. Man had drilled and dredged and dammed here but the marsh still looked and felt wild—the scuttling fiddler crabs, wind sighing through the cattails, and the vast flocks of waterbirds dropping onto the wavy grass like snowflakes on a meadow.

The sun was about to punch out for another day and the sky was vermilion below stringy blue clouds. Below me the mosquitoes were tuning up like the string section of an orchestra before a concert. I focussed the binoculars on the setting sun and transformed it into a great orange dirigible subsiding onto the horizon. A dozen ibises glided across the midsection of the dome-shaped sun and seemed to freeze there in silhouette for several seconds. Then it was over, the sun was gone, and the mosquitoes were forming up for a twilight attack.

A chain of more than a hundred coral reefs spangles the sea along the Florida Keys to the Dry Tortugas 65 miles beyond Key West. This living reef system makes a vivid and dramatic underwater landscape, rising in ridges and cliffs like a miniature mountain range complete with canyons, draws, and caves. The varying coral shapes are produced by different types of polyps, the tiny creatures that attach to a submarine surface and secrete the hard, limy material that builds into a reef.

Here along the keys, the gulf waters surge against those of the Atlantic and have created a navigational obstacle course for generations of seamen. Luckless ships with names like *Triumph, Ajax* and *Tennessee* have sunk to the bottom in these waters. Now modern-day voyagers can get a safer look at this stretch of coastline at John Pennekamp Coral Reef State Park, a 21-mile-long by three-mile-wide band of turquoise water off Key Largo. ("Key" comes from the Spanish *cayo,* or islet.)

Queen angelfish cruise Florida's colorful Pennekamp coral reef (left). Myriad shapes such as the delicate sea fan (right) are formed by secretions from polyps.

The most popular diversions among the 600,000 people who pass through Pennekamp yearly are snorkeling and riding the glass-bottom boats that cruise the reef. A fortunate few can ride in the company of a park biologist fittingly named Jeanne Parks, a sunny-tempered fugitive from the Midwest on her way to becoming a "conch" (pronounced *konk*), the shellfish name by which oldtimers in the keys are known.

As the boat chugged toward Molasses Reef eight miles offshore, the water color changed subtly from green-gray to blue-green to light blue. The lighter patches, Parks explained, mean a sandy bottom. Dark water indicates seagrass-bottomed shallows. The water at Molasses Reef was a rich blue.

The tiny coral polyps that build the reefs have four non-negotiable requirements that the Florida Keys supply in abundance: salt water, a good food supply, warm temperatures (68° to 86° F), and water so shallow and clear that sunlight can reach the algae inside the polyps.

My first impression of the reef was sensory

A bird's-eye view of the Florida Keys shows masses of coral reefs (far left). From anchored boats, snorkelers plunge into the warm water to admire formations such as staghorn coral (left).

overkill, a rush of colors and forms and movement that took a few minutes to assimilate. The dominant color was a blend of olive green, light brown, and pale mustard, reflecting the algae coloration in this subtropical habitat.

The coral polyp has the ability to secrete lime, creating a cup around itself that is as hard as rock. The patterns (stars, flowers and so forth) are created by different kinds of polyps.

The names we attach to different coral shapes derive, of course, from our own experi-

ence. Elkhorn and staghorn coral look exactly like their names, only larger. At places the floor of the reef was piled high with so many dead, antlerlike deposits that it resembled a vast elk shedding-ground.

The round, boulderlike deposits called brain coral—possibly a premed student's inspiration—contained little grooves reminiscent of the canals in closeup pictures of the moon. Sea fans, sea plumes and whips—all flexible types of coral that do not build reefs—oscillated gently. The fans looked like brown elephant ears.

Fish in a score of colors and a dozen sizes glided amid the mounds of coral. A blue parrotfish with a splash of yellow on its side nibbled at algae on a shard of living coral—for a parrotfish the reef is both home and cafeteria. Altogether, parrotfish scrape about a ton of coral off each acre of reef annually in the process of feeding on the algae. Some parrotfish secrete, in effect, their own blanket, a thin, veil-like mucus layer that envelops them as they sleep. The mucus protects them from predators, but exactly how is not known. It may disguise the fish or may even chemically repel predators.

Shiny French angelfish, their steel-gray bodies dotted with hundreds of bright yellow punctuation marks, mingled with platoons of smaller, yellow-and-black-striped sergeant majors, perhaps so named by a retired noncom turned snorkeler.

Parks pointed out a foureye butterflyfish, whose false eye markings on its backside confuse predators and draw them away from its head. "And there's one of my favorites," she said, indicating a dark, six-inch-long fish that appeared undistinguished except for a bright, butterfly-shaped yellow tail. "That's a damselfish, feisty as they come," she said. "They attack larger fish out of territoriality."

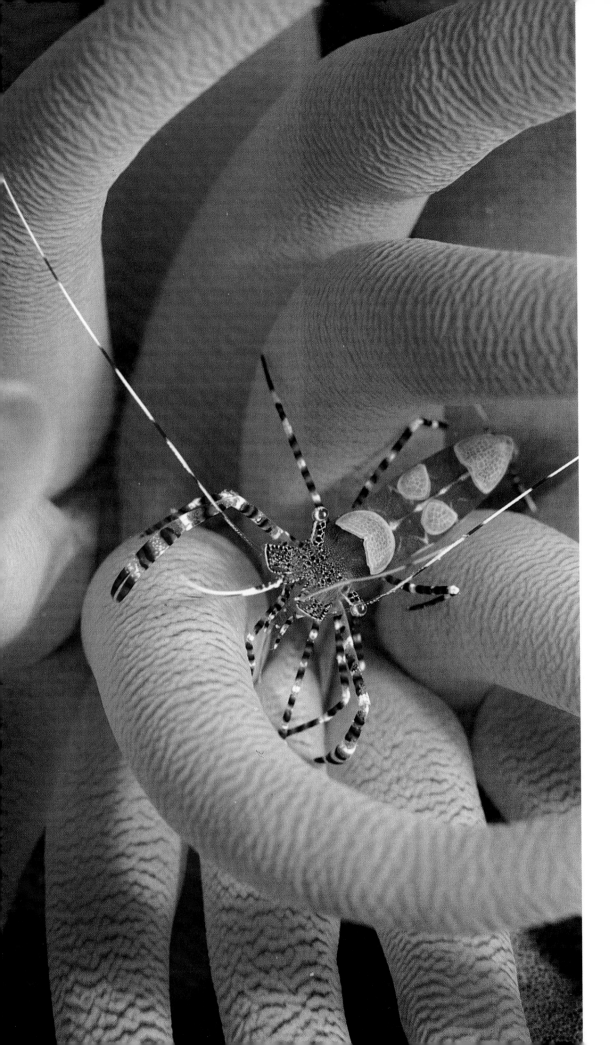

The spotted cleaner shrimp moves with impunity among the paralyzing tentacles of anemones. The shrimp feeds on parasites it plucks from the scales and mouths of fish.

Fish cruise the reef in astronomical numbers because of the plentitude of living morsels there—plankton and worms, minnows and shrimp and crabs. It's the numbers that make the reef so spectacular, so many creatures swirling about that you almost feel claustrophobic. It's not unusual to see a hundred or more fish of the same type crowded together, waiting for the dinner bell. Reefs have day and night shifts, the daytime feeders retiring to recesses and hollows amid the coral when the night crew comes on: squirrelfish, bigeyes, moray eels and octopuses.

Many reef-dwellers develop mutual-aid strategies, which scientists call symbiosis. Coral shrimp, for example, emit a signal that certain fish understand as meaning the cleaning-station is open for business. Fish line up to be picked clean of parasites by the shrimp; the fish get a grooming and the shrimp get a meal.

Sharks show up infrequently at reefs, but tour guide Milli Hadley recalled one trip when a 30-foot-long whale shark suddenly appeared, filling half the starboard windows. "It was like a dream," she said. "It lasted about 30 seconds and the passengers were so stunned they didn't even *ooh* and *aah*."

Pointing out an unimpressive-looking little fish called a wrasse, Hadley told the group that when a male wrasse dies, the strongest female in the school turns into a male within a couple of weeks. And a male wrasse, she added, saving her spiciest tidbit until last, is male indeed, able to mate as many as 40 times a day.

After the boat returned to the dock, Parks led the way to another phenomenon characteristic of the keys, an elevation known as a "hammock." Five to ten feet higher than the adjoining land, hammocks are clogged with trees whose wood is heavier per cubic inch than other types. This adaptation helps them survive hurricanes.

"Even in hundred-mile-an-hour winds they'll lose twigs but not branches," said Parks. When a hurricane makes a direct hit on a reef, however, it rips coral formations and smothers them in sand, killing fish outright and destroying their habitat.

For all-around handiness the most impressive hammock tree is the musically-named gumbo limbo. This copper-trunked tree boasts the remarkable ability to regenerate from its branches. Break off a branch and stick it in the ground, and it will develop its own root system and keep growing. In a tropical climate where dead-wood fences quickly rot, this is a commendable attribute.

Setting out on a hammock trail walk, Parks paused to scoop up a handful of what passes for soil in the keys. "It's dried-out leaf dust," she said. "They call it 'duff.'" Soil as we commonly know it doesn't develop here because the growth cycle never slows enough to permit it. The shallow-floored hammock habitat supports several types of land crabs and snakes along with raccoons, Eastern gray squirrels and the spectacular, yellow-winged swallowtail butterfly.

"Oh, wow, look at this," she cried abruptly, pointing at a wingless, bright red wasp moving placidly across the trail. It turned out to be a wasp with an outsized reputation, a female

velvet-ant wasp. Its favorite gambit, Parks said, is to lay its eggs in the nest of another species of wasp. When the velvet-ant wasp larvae hatch they consume the other wasp's eggs along with the spiders the resident wasp had stored as food. Velvet-ant wasps have a sting like a scorpion and a shell like lead casing. "I stepped on one once and ground it under my boot," Parks said. "It just crawled away."

The reefs and the great colonies of wildlife that live on and around them confront threats such as pesticide runoff and the possibility of oil drilling near Pennekamp's borders. It helps immensely that the rules nowadays regulate boaters in the reef area and forbid touching the delicate coral, but it was not always so.

This unusual underwater state park grew out of an alliance between two men, a scientist and a journalist, who loved the unique beauty of the reefs and realized that the colorful formations were in danger of being hacked to pieces.

Gilbert Voss, a marine biologist at the University of Miami, first called attention to the appalling reef destruction in the 1950s. On one weekend in 1957, Voss discovered more than 50 boats, including a sizable barge, collecting coral chunks to sell as souvenirs. That same weekend, treasure hunters were dynamiting a coral deposit to get at a Spanish wreck while several divers manhandled fragile coral growths to flush tropical fish from their hiding places.

Voss proposed a park, and soon gained a key ally in *Miami Herald* assistant editor John Pennekamp. A longtime conservationist who had helped establish Everglades National Park, Pennekamp plumped for the idea in his newspaper column and led the three-year campaign that resulted in the park that was named for him.

The best way to experience the reef is by scuba diving or snorkeling. For this reason I suppressed my apprehension as a C-minus swimmer and joined a snorkel trip to the same Molasses Reef we had seen from the boat. "If you get sick," the laid-back skipper cheerfully advised the company, "we'll just drag you behind the boat on the way home."

When we reached the reef I donned my equipment and plunged in. The sound struck me first. It was a faint, distant cracking, like popcorn popping two rooms away. I was later told the noise was probably thousands of snapping shrimps, also known as pistol shrimps. The subsurface color was bluer and more vivid than it had looked from the boat, and the sense of swarming, milling masses of fish was more overpowering. Hundreds of wrasses, the sexual superstars we had seen earlier, idled in the lee

of a coral cliff; large schools of the gray fish inelegantly named grunts swam toward me and gave way grudgingly, one bumping my mask as it swerved aside.

About 30 feet away, a half-dozen barracudas cruised slowly in a wide circle. Other reef fish glided or darted modestly, even diffidently, but not barracudas—if a fish can be said to swagger, the barracudas swaggered. They appeared to be the bullies on the block. But even the presence of barracudas couldn't break the mood of otherworldly peace that ruled here. It pervaded the landscape of canyons and ravines, and gave this world of marine creatures and heavenly blue water a tranquillity not matched on the surface.

The first explorers to cruise the southwest coast of Florida must have gaped in astonishment when they saw the shoreline now known as Ten Thousand Islands. What they beheld was not sandy beach or dunes or rock cliffs but trees—trees that seemed to emerge directly from the water, in great coastal forests and tiny two-tree islets, trees as far as the eye could see. These were unlike any they had seen before, curiosities with their trunks perched atop a tangle of roots that grew above the water in dense thickets. The Indians called them "walking trees" because they appeared to walk on water. We call them mangroves.

Mangroves crowd tropical coasts in a broccoli-like mat. Three varieties—red, black and white mangroves—thrive in Florida because their biology enables them to tolerate salt but not frost. They either block salt's entry at the roots or excrete it through glands on the leaves. Mangroves have also found a way to

Salt water kills most trees, but mangroves (left) thrive in it, their exposed roots forming a tangled mat. Herons (right) comb the mangrove shores for snails, insects, and fish.

survive in muddy ooze almost devoid of oxygen. They manage this by "breathing" through pores in their above-ground roots or in little projections that poke through the brown muck like fingers pointing skyward.

Mangroves create a rich habitat for aquatic life because of the organic material—dead leaves and sediment—that gathers around their roots. For marine creatures a mangrove root system is an apartment house and mall in one complex—it has everything they need. Leaf particles and the bacteria they carry begin a food chain that rises through oysters and crabs and fish to herons, ospreys, and eagles. The oysters attached to roots exposed at low tide confused the English explorer Walter Raleigh, who reported that in the New World the tasty mollusks grew on trees.

The most imposing submarine resident of the dark channels and shallow inlets that flow amid the Ten Thousand Islands (a number that skeptics calculate as too high by about 9,500) is the manatee. Protected in Florida for almost a hundred years, leather-skinned manatees look like oversized walruses (their average weight is

Placid and passive, the endangered manatee (right) propels through channels and canals by undulating its hind end and steering with its flippers and tail.

Anchored underwater, mangrove roots form an elaborate thicket. Organic material that accumulates around the submerged roots is a rich source of food.

1,000 pounds) or mild-mannered hippos with flippers. Commonly known as the sea cow, the manatee is one of the few mammals that spend their entire lives in the water. About a third of that time is spent eating; a full-grown manatee can eat 100 pounds of seagrass a day.

Manatees are small-brained, unaggressive animals often seen touching and nuzzling in apparent affection. Their main enemies are cold water—they stop feeding at 60° F—and large power boats, which are responsible for most manatee deaths.

Although manatees generally come in peace, a surprise encounter with one can be memorably unnerving. Paul Purifoy, a ranger at Everglades National Park (most of the Ten Thousand Islands region is in the park), was patrolling one day when a manatee suddenly exploded through the surface a few feet from his boat. "This enormous tail flew out of the water and drenched me with dirty water," he recalls. "I practically jumped out of my skin."

With few exceptions, people tend to shift into reverse when they approach a mangrove shore. The reason is that mangrove shores stink.

The agile snowy egret moves easily through the roots of a red mangrove. The tangled maze provides safety and a place to find meals of frogs and snakes.

Low tide exposes oysters clinging to mangrove roots (right). Mangroves' dense foliage makes a desirable nesting spot for birds. During courtship, the male yellow-crowned night heron (below) presents nesting material to the female.

In warm weather, especially when the air circulation is poor, the mangrove shores are full of the odor of hydrogen sulfide, which indicates a shortage of oxygen. They're also oppressively humid, full of mosquitoes and impossible to traverse afoot. Since we didn't find much good about them until recently, humans have been inclined to hack mangrove trees up for firewood and to bulldoze and drain the swamps for developments like Marco Island, just north of Ten Thousand Islands.

But with the recognition of the role mangrove coasts play as aquatic nurseries, they have gained overdue protection. Permission is now required to cut down a mangrove tree in Florida; in some places anyone who fells a mangrove is obliged to plant a new one.

Historically, the only people who found anything positive about mangroves were the hermits, fugitives and other maladapted sorts who lived in Ten Thousand Islands early in this century. One legendary islander named John Gomez claimed to be 106 years old when he brought his 78-year-old bride to his palmetto thatch hut on Panther Bay. He liked to recall his career as a soldier/pirate in the gulf, and a blockade runner during the Civil War. The white-bearded Gomez was 122 by his reckoning when he died in 1900—by drowning.

Ted Below, white-bearded like Gomez was, is a reconstructed New Yorker who now serves as warden of the National Audubon Society's Rookery Bay sanctuary near Marco Island. Twice a month he slowly cruises the preserve's mangrove-fringed waters to monitor the water-bird population. Below *wants* people to think ill of mangroves—"so they'll stay away." His stories are mostly about birds, especially brown pelicans, and about human folly in various forms. He refers affectionately to the Gulf of Mexico as "that lady out there." When residents of Marco Island proposed dredging a sandbar, for example, Below's reaction was, "that lady out there will put the beach where she wants to."

On this bright October day he was making his regular bird-counting run and a side trip to Ten Thousand Islands. As Below's boat buzzed across sunlight-dappled Rookery Bay, a half-dozen mangrove isles from 100 to 500 yards wide protruded from the water. Herons and gulls and two pink-winged roseate spoonbills congregated on a barely visible bar.

Spoonbills are the only pink wading bird native to the United States. They are exclusively Gulf Coast dwellers, with the Florida delegation partial to mangroves as nesting habitat.

Spoonbills appear to have been color-coordinated by a practical joker. Their pink body plumage clashes with their bright orange tails, red eyes and a green bill that turns gold in mating season. The fiery color of their plumage is created partly from a diet rich in carotenes. In the muddy water, they feed by feel, sweeping their spatula-shaped bills until nerves on the bill edge detect food. They are not interested in seeing their vittles; in fact, a spoonbill in captivity will ignore shrimp placed in plain view in front of it. Only when the bill touches the shrimp does the bill twitch into action.

Several egrets, the first birds to come in when the tide falls, settled in the shallows near a sandbar. In the muck a solitary black-and-white oystercatcher probed for food. The red bill of an oystercatcher is as functionally specific as the spoonbill's—flat, sharp and strong, and used like we use a knife to open a mollusk's shell.

As the boat neared a tennis-court-sized mangrove island, Below pointed to several dozen brown pelicans clustered close together in the branches. In his work Below focuses on brown pelicans, classified as a species of special concern in Florida, as an "indicator" species reflecting the vitality of the habitat. Below has documented a four percent annual reduction in pelican numbers at Rookery Bay for the past seven years, in part because engineers have diverted much of Florida's fresh water flow to populated areas.

Below has watched the roughly 500 pairs that nest around Rookery Bay so long that the birds are like family to him.

When a female pelican joins a male at a nest site selected by the male, "he drives her off until he's sure she's serious," Below said. Once the touchy male is satisfied he collects nest material and the female begins construction.

Below enjoys studying pelicans "because they're always making a liar out of you. They were thought to be silent except for the chicks. But one day I was out here with a photographer when a pelican suddenly squawked—so much for the silence theory." It was also believed that brown pelicans never completely submerge during their dramatic, wings-back dive into the surf for fish. "So right in front of me one day several dive and stay under for 30 seconds. They're fascinating but frustratingly unpredictable."

He maneuvered the boat between a narrow sandbar and the slender beach fringing

Marco Island, with its row of blocky high-rise buildings. A pelican smacked into the surf nearby at a 60-degree angle, churning up a miniature geyser. "Wait," Below cried suddenly, snatching up his binoculars and zeroing in on a smallish, sand-and-white bird. "I have to see if that piping plover is one of mine." A feisty political operator, Below worked hard to get part of this sandbar protected as a "critical wildlife area" for the piping plover, a threatened species in Florida. Thousands of them now winter here.

Plovers feed on the run, sprinting short bursts to snare insects and small crustaceans. A piping plover—so named for its shrill, whistle-like two-note call—is born with the ability to scoot: a chick only two days old was once clocked at 4 m.p.h.—not racehorse speed, to be sure, but not bad for two days old.

Below shoved his boat into a full-throttle run across 10 miles of open gulf, ending at the northern edge of Ten Thousand Islands. On the horizon dozens of islands in all sizes and shapes fused together. A line of mangroves shows how the trees colonize land: sediment deposits collecting around the roots reach out to connect each tree to its neighbor five yards away. A system of branched roots anchors the trees, holding them steady as tides move in, then out. Nearest to the sea is the red mangrove, towering as high as 40 feet. Closer to land is the scrubby black mangrove, then the white mangrove.

An osprey rose inquisitively from its nest on the highest tree in sight. To waterbirds, the mangrove shores—full of insects and snails and little fish—are a delicatessen. Below bumped the boat onto the beach at Panther Key, stepped nimbly on a half-submerged mangrove trunk and plunged into a thicket of roots, leaves and low branches. Many of the leaves were yellow-brown because mangroves shed throughout the year. Healthy mangroves can drop about four tons of leaves per acre each year. Each fallen leaf is attacked by bacteria, in six months decomposing into a feast for thousands of bacteria and microrganisms, which feed worms and crustaceans. In turn, these are food for crabs, shrimp, and beetles, which feed herons, fish, and spoonbills. The tangle of mud and roots may be easy navigating for them, but not for humans. Fifty feet of mangrove-trekking was enough for Below that day. The raccoons, rabbits, snakes and other residents did not wish to socialize at this time. The mosquitoes, however, appreciated the company.

Toward sunset Below cut the motor. The boat drifted into a pool beneath a canopy of mangrove branches on the small island where we had seen the pelican roosts earlier. Homecoming egrets plopped onto the branches like tired commuters; smaller birds—cattle egrets and little blue herons—roosted in the interior trees. The island was an avian housing project; by dusk thousands of birds would be there.

Below sniffed the fresh-smelling air and cocked an ear to the soft chatter coming from the nests of birds. "Ah," he sighed, grinning, "so this is the horrible, dank mangrove forest with all the creepy-crawlies." Wispy shafts of light stutter-stepped through the branches. Pelicans circled in a light breeze overhead.

A young pelican, no longer a nestling but not quite independent either, began squawking insistently. "He wants food but he won't get it; it's too late," Below said. At that precise moment the mother pelican arrived with a fish. Below smiled and shrugged. "They just made a liar out of me again," he said.

Introduced from the West Indies, the Cuban tree frog (right) is now common in southern Florida. Large toe discs help it cling to the branches of mangroves.

PACIFIC

California sea otter

Everyone can picture what an American Pacific Coast beach looks like. It's a sloping carpet of powdery sand perpetually bathed in sunlight. It's located in southern California—somehow northern California, Oregon, and Washington are in another world—and it has a name like Balboa or Malibu. Large waves called "combers" roll to the beach looking as perfect as Frankie Avalon's wavy hair, while bronzed young men skim atop the waves on surfboards. Beach bunnies multiply on the sand, flouncing around in bikinis that will never keep the textile industry in business. In the background frolic carefree kids riding inner tubes and teenagers playing volleyball. It's heaven without the clouds.

At least, that's the image promulgated by soft-drink ads and 1960s beach-party movies. Even if you scoff at this image of Pacific Coast beaches, don't dismiss it entirely, for it contains an element of truth: southern California beaches are indeed thought of as playgrounds. The perception of a sandy beach as a fascinating ecosystem gets lost amid the beach towels and hot-dog stands.

A southern California boy myself, I grew up frequenting those golden beaches. It wasn't until my teenage years that I had

Sandy beaches lure thousands of people to sun and surf. More than a place to relax and play, however, beaches also host an often unnoticed world of fascinating creatures.

my first memorable (though accidental) encounter with the beach as a natural environment. I was in a group of teens staying up late to witness a grunion run, but I didn't know grunion from goulash. All I knew was that if Mary Helen somethingorother (I can't even remember her last name now) was going to watch grunion, so was I.

At about two in the morning we stationed ourselves just above the high-tide line. To my surprise, hundreds of other people were strung along the beach too. Though I was hungering for a crumb of attention from Mary Helen, a spark of scientific curiosity flickered in my preoccupied mind. What was going on here?

Then the first grunion bodysurfed ashore and slithered up the cool, wet sand. Others followed. The silvery fish thrashed like … well, like fish out of water in their attempt to get as high up the beach as possible. When they could struggle no farther, the females drilled tailfirst into the sand, burying about half of their six-inch bodies. Quickly several males wrapped around each female like crescent moons, fertilizing the eggs they were laying. Seconds later the next wave swooshed up and the males rode it back to the sea. A couple of waves later the females followed. Subsequent waves brought in more grunion than I could count.

A sense of wonderment stirred in me that night. Where did the grunion come from? How had everyone known that the grunion would show up on this beach at this time? Excited by the revelation that there was more to the beach than bodysurfing and body watching, I forgot all about Mary Helen.

Perhaps it's understandable that it took me so many years to perceive the beach as a natural system as well as a playground. Finding a pristine sandy beach in southern California is nearly as hard as finding a parking space. Yet even those people lucky enough to stroll a beach in solitude can miss much of nature that is hidden.

That fact was brought home to me again when I recently walked the beaches of Oregon Dunes National Recreation Area. Its 41 miles of shoreline constitute the longest undeveloped sandy stretch on America's Pacific Coast. No better place exists in which to contemplate the minimalist landscape of sand, sea, and sky. Consider the smooth surface of the water on a calm morning before the sun rouses the ocean's blue: it is more the essence of a gleam than any

Californians line the beach in the middle of the night to watch grunion flop up the sand to lay eggs (above). In a precisely timed ritual, the female buries her tail in the sand (right), then several males deposit their milt next to her.

color. Breathe in the salty scent of the sea—it's as much taste as smell. Listen as the surf sings its simple, repetitive song, a chant that is somehow thunderous and muffled at the same time.

The first sign of life on Oregon Dunes' beaches is the wrack, or things washed ashore from the open ocean. Scattered among the debris are manmade objects such as fishing lures, bottles, and crates. Traditionally, beachcombers covet the colorful glass floats that drift in from Japanese and Soviet fishing nets, but these days the floats are more likely to be made of plastic.

Even glass floats pale in comparison to a message in a bottle. Happily, shipwrecked sailors are in short supply, but oceanographers re-

The tough little beach hopper shifts with the tides, constantly moving to stay in moist areas. At night, the beach may be thick with hoppers munching on morsels of washed-up seaweed.

lease drift bottles by the thousands. Inside are cards that the people who find the bottles return to researchers, telling them when and where the bottle was found. That allows scientists to plot the movement of surface waters. In return, oceanographers send letters to the finders telling each of them when and where his bottle had been set adrift.

More abundant than manmade debris is the natural wrack. Strolling Pacific beaches, I've seen everything from the tiny, nearly transparent comb jellies called "cat's eyes" to a whale skull the size of a washing machine. In just one day I saw a dead ray, the body of a radiantly amber jellyfish, the carcasses of at least a dozen juvenile common murres that had flunked natural selection, and a small shark.

Even the foam that piles up along the beach is actually life in disguise. Common in spring and summer, this bubbly mass is really billions of single-celled plant skeletons that have been glued together by their cell sap and have trapped air in the spaces among them.

Perhaps the most wondrous wrack resident is the "by-the-wind sailor." Sometimes after a west wind in the spring, millions of these small, purplish jellies wash up on a beach, making the beach itself appear purple. But what fascinates beachcombers is the by-the-wind sailor's method of locomotion. Atop the creature's flat, roughly oval body stands a triangular sail set at an angle to the sailor's major axis. This allows by-the-wind sailors to deviate from a straight downwind direction by as much as 60 degrees. Why? No one knows for sure. Also puzzling is the fact that half of them are "left-handed" and half "right-handed," meaning their sails are canted in opposite directions. Consequently, half of them sail only to the left of downwind and half only to the right.

Wrack has one shortcoming as the beach's first sign of life—it generally consists of things that are dead, dying, or never were alive. That's because life on a sandy beach is subject to the edicts of the tide, to relentless body blows delivered by the breakers, to scathing winds and summer heat, and to the appetites of winter storms that can devour entire beaches overnight. Even more treacherous is the incessant shifting of the sand—hardly what you'd call a

stable home environment. To cope, most animals burrow in the sand or engage in a never-ending dance with the tide.

Beach hoppers—often called "sand fleas" —usually bury themselves in moist sand during the day. But if you happen to walk the right beach after dark with a flashlight you may come upon a beach-hopper jamboree. The hoppers

Sea winds push most floating things straight ahead, but by-the-wind sailors (above) move diagonally, like sailboats on a reach. This is because each creature has a triangular "sail" standing up at an angle to its body (left).

bound about like jackrabbits, occasionally pausing to sample the seaweed smorgasbord. The standout among Pacific Coast hoppers is the mammoth California beach hopper, which, bright orange antennae included, can measure a strapping two-and-a-half inches tall.

Like beach hoppers, mole crabs move in tune with the wave action, but their need is to stay *in* the wash of the waves, not out of it. They generally dig themselves into the sand and face the ocean with nothing showing except their first pair of antennae, which serve as breathing tubes, and their stalk-mounted eyes, which protrude from the sand like tiny periscopes.

As a spent wave recedes over them, mole crabs unfurl their second pair of antennae. Shaped like the plumes on a Victorian lady's hat, these catch the minute animal and plant matter streaming past. Dining in such a maelstrom car-

ries certain risks, but mole crabs are outstanding diggers—hence their name—and when wave action knocks them loose, they burrow frantically back into the sand.

To get a feel, literally, for how they dig, scoop up a handful of sand with a mole crab in it. In a blink it will bore through the inch of sand to your palm and start clawing furiously at your skin. It doesn't hurt, but the ferocity of the effort often startles people.

Small creatures like mole crabs eat mainly what gets washed in, from microscopic plankton to globs of seaweed. Bigger creatures of the beach, in turn, eat the many small creatures. Even in the few inches of sand under the surface, predators abound—in this case, worms. One such prowling carnivore is *Cerebratulus californiensis,* which has not yet been tamed with a common name. This worm, burnt-orange and ribbon-flat, grows to lengths of a foot or more. It seizes its prey with an enormous proboscis which is covered with sticky mucus. Eugene Kozloff, a University of Washington zoology professor, writes that this mucus is so adhesive that when he let such a worm wrap its

proboscis around his finger, he had to peel it off like a strip of tape.

Not even Kozloff would care to tangle with *Nephtys californiensis,* one of the other worms that dig and hunt in the sand. This iridescent worm can inflict painful wounds with its short, stubby snout.

While rambling near the Oregon Dunes one day, a marine biologist and I shoveled up what we thought was a champion specimen of *N. californiensis.* Another scientist took a look at our find and shook his head disparagingly. When we asked how large a specimen *he* had encountered, he held his hands about two feet apart and then made a circle of his thumb and index finger to indicate the worm's thickness. Still want to stroll down the beach barefoot?

These worms and their brethren aren't the only predators at the beach, however. When the tide is in, sea stars and crabs move up from the depths to feed. The long bills of shorebirds are always probing for some unlucky creature. If you happen to be a clam, though, one predator stands out above the rest, and that is the person wearing hip boots and brandishing a shovel.

Though most of the clams we eat prosper in the sand and mud of bays, estuaries, and other protected areas, the open sandy beach does provide a home for two esteemed varieties: the razor clam and the Pismo clam.

Razor clams rely entirely on digging speed to survive in the melee of the surf. When unearthed by a wave they are able to bury themselves completely in just a few seconds, before the next breaker strikes. A razor clam digs with its foot, an organ that can protrude half the length of the shell. The clam thrusts the slightly hooked tip of its foot into the sand, swells the tip until it balloons into an anchor, then contracts the muscular foot to haul itself down.

To capture this most elusive of Pacific clams, expert advice is called for, so I called for Terry Link. Link has been chasing razor clams since he was a kid, and now he's a shellfish biologist and the man in charge of Oregon's razor clam fishery. "Digging razor clams is an art," he says. "You don't want to get discouraged after just a few times." According to Link, the first step is to go out at the right tide in the right weather on the right beach. Then you've got to home in on what veteran clammers call "show," subtle signs that indicate a razor clam's presence beneath the sand. Show can be a dimple the size of a dime, a minute "V" in the backwash as water splits around a clam's slightly protruding siphon or, if you're searching in shallow water, the slight depression that a razor clam makes when it jerks down its siphon at your approach. Often all you'll see is "a faint, shallow hole that could be anything," as Link encouragingly puts it.

Armed with digging tools, keen eyes and lots of patience, clam diggers descend on Pacific shores just after dawn in hopes of bagging tasty rewards.

Locating a razor clam is just the beginning. Link recommends that you next insert your shovel four inches to the seaward side of the show (because razors will dig in that direction), remove two or three shovel loads of sand without pivoting the shovel (that could break the fragile clam shell), then grope—gently—with your hand into the slush until you feel the clam. Gently, because, as Link says, "If you jam your hand in and you've broken the shell, you'll find out why they're called razor clams." You might consider buying the condensed cream of potato soup and letting someone else get the clams.

The first time I set out after razor clams, I

simply shoveled a bladeful of sand onto the beach, and there was my first razor clam. Not so tough. Maybe the warnings about the razor's elusiveness and speed had been exaggerated. The clam was smaller than the legal limit, so I turned away to continue my quest. Then it occurred to me that I should see just how fast a razor clam really can dig, so I turned back to look . . . but there was nothing to look at. The clam had quickly burrowed away.

I didn't catch another clam that first outing, but just being out on the beach at dawn was enough. In the rumpled blanket of fog, phantom figures bearing shovels and nets continued working the darkly shining sands like so many hungry shorebirds: it wasn't the ambiance that drew these clammers to the shore, however.

Sport diggers have taken millions of clams on the Pacific Coast. Nowhere is this better illustrated than in the Pismo Beach area in central California, where generations of Americans have bagged famously tasty Pismo clams. In the late 1800s, farmers plowed up the clams by the wagonload, feeding the seemingly inexhaustible supply of shellfish to their stock. Once 150,000 people made off with 75,000 pounds of clams in a single weekend. Not surprisingly, such gluttony couldn't go on forever. Today it's difficult to find even one Pismo clam of legal size on Pismo Beach. Some clammers blame otters (or moon snails, or crabs, or shorebirds) for reducing the number of clams, but history suggests that people share some of the blame.

It would be a shame if the Pismo clam disappeared from the beach, because the large, strong-shelled Pismo is an excellent example of how creatures have adapted to the shore. This clam doesn't just survive in violent surf; it *needs* heavy surf. When removed from the high oxygen content of turbulent water, a Pismo clam will soon die. When left where it belongs, however, a Pismo can live 40 to 50 years.

The breaking waves also mark the beginning of the domain of ocean dwellers—harbor seals, whales, dolphins and the like. Along Oregon Dunes' beaches it's not uncommon to be taken aback by the sight of a whiskered face peering from the shimmering arc of a breaking wave: a harbor seal chasing perch and flounder in the surf. Sometimes harbor seals haul out on

the beach, but even then observers get little more than a tease because the seals flop back into the water at the first sight of a human.

Northern elephant seals are another matter. They come ashore for much of the winter to breed and give birth. They return to shore again in early spring and summer to molt. During the 19th century, these enormous pinnipeds were hunted nearly to extinction for their blubber, but since protection and reserves were established, the seals have made a comeback.

Today Año Nuevo State Reserve in central California hosts the largest mainland northern elephant seal colony in the world. Bulls battling over breeding rights put on quite a show, and in 1973 a popular regional magazine sang the praises of Año Nuevo's wildlife spectacle. Enthused but ignorant, thousands of people invaded the beach. Some trampled the reserve's

Close quarters don't seem to bother elephant seals (left), which bunch up in colonies. The gregarious creatures come to sandy shores to molt, mate and give birth to pups (above).

fragile dunes, others ran around in the midst of the bewildered seals. After some quick training, students from nearby colleges helped protect the seals from overzealous visitors and led interpretive tours of the colony.

Today a delicate balance exists at Año Nuevo between the needs of the elephant seals and the wants of the people who come to see them. About 3,000 elephant seals occupy the mainland colony during the winter, 1,000 of them new pups, and about 50,000 people come to watch. During the winter, only people on guided walks are allowed in, and though 27 walks are given each day, weekends are booked up the first day reservations are taken.

Looking at an adult male elephant seal from the back, one might assume seals got their name from their awesome bulk—lengths up to 16 feet and weights up to 8,000 pounds. Look at an elephant seal head on, however, and the reason for its name will be as plain as the nose on its face. A bull is burdened—or blessed—with a colossal inflatable nasal sac. Shaped like a wrinkled boxing glove, this snout droops almost into

a bull's mouth, putting it into a league with anteaters, elephants, and Jimmy Durante when it comes to preposterous proboscises.

The snout can swell and straighten like a crumpled paper bag blown full of air, as I discovered years ago when I came across an elephant seal colony on a remote Pacific beach. As I watched the 25 or 30 animals going about the business of mating, nursing, and playing, the beachmaster—the dominant male—suddenly surfaced from the water about 50 feet from the rock on which I sat. Apparently he considered me an intruder, perhaps a rival for his harem (not my type, actually), because he swam slowly back and forth strutting his inflated snout and emitting spluttering, gurgling noises.

This three-ton sausage seemed laughable until a genuine rival, another male elephant seal, crept onto the end of the cove. Instantly the beachmaster erupted from the water and took off down the sand at an earthshaking gallop. The intruder took one look at this onrushing steamroller and fled. Triumphant, the beachmaster returned to the water. Like a victorious boxer bowing to all four sides of an arena, he revolved in quarter turns, sounding his bathtub-drain gurgle at each turn.

Few creatures are larger than a bull elephant seal, and the only one of these likely to be spotted from the beach is a whale. Most whales appear rarely and randomly, but gray whales often pass close to the shore with clockwork regularity. John Goold, a wildlife biologist at Oregon Dunes, reports that he has seen gray whales cruising past the area's beaches less than 100 yards away. An estimated 21,000 gray whales migrate along the coast between their feeding grounds in Alaska and their breeding grounds in the lagoons of Baja California, attracting thousands of whale-watchers.

During the three or four months of the gray-whale migration, the overlook at Cabrillo National Monument near San Diego draws enthusiastic whale-watchers by the thousands. Atop Cabrillo's bluffs on a winter weekend, about 3,500 whale-watchers crowd along the clifftop and walk over the life-sized gray whale drawn on the lookout's blacktop. Ranger Roger Covalt watches many people who have never seen a whale before. "Their eyes open wide and they get incredibly excited," he says. "It's like when you buy your first new car."

Away from the crowds, standing on a remote beach on a misty morning, the experience of seeing a whale can elevate from exciting to mythic. A 10-foot spout of water draws your eye to the surfacing whale, which might be so close that you can see barnacles on its back. The wet-shiny, black-gray body emerges as the water is still raining back into the sea. Perhaps 50 feet and 40 tons of whale slide by before the great beast disappears back into the ocean.

Such sights almost involuntarily turn our gaze seaward. But if you're strolling on many Pacific Coast beaches it's worthwhile to look inland as well, for there lie the dunes, an extension of the sandy beach.

Pacific Coast dunes shoulder up to tremendous heights compared to their Atlantic cousins—500 feet compared to about 30. But a beachcomber's view of the monster dunes, which sometimes sit back a mile from the ocean, may be cut off by the vegetated foredune.

In some places the foredunes cut off more than the line of sight. For example, in Oregon Dunes, the intrusion of European beach grass has enabled the foredune to grow much taller than is natural, blocking the wind-blown beach sand that builds the dunes.

For now the dunes are still plenty high, as you will discover emphatically if you struggle up one using the traditional two-steps-up/one-step-backslide method of ascent. It's worth the effort, though, especially for the views. Much of Oregon Dunes has a monochromatic, *Lawrence of Arabia* look, and in fact, numerous "desert" movies have been filmed there. But the vista from a 500-foot summit will reveal that the dune fields are a complex ecosystem in which bare dunes share space with a thickly vegetated plain, grass hummocks, and oases of forest.

Here and there bursts of color tell of wildflowers, such as coast morning glory, seashore lupine, and king's gentian. Truly wonderful are the perennial streams that appear, like mirages, from the sides of barren dunes and run only a few yards before disappearing back into the sand. Sometimes yellow coast monkey flowers brighten these improbable watercourses.

The vegetated areas and the bordering forests brim with life, as evidenced by the multitude of animal tracks left in the sand overnight. Lie prone atop a high dune on a full-moon night and you're liable to witness the comings and goings of mice, raccoons, deer, owls, and skunks. Coyotes sometimes fill the profound quiet with their yips and howls.

Next page: In vast sand dunes along the Oregon coast, sturdy grasses and spruce trees nestle along with delicate wildflowers. After dusk, raccoons, coyotes, and bobcats leave behind a crisscross of tracks for morning visitors to unravel.

Even if no animals appear, the dunes reward a twilight sojourner. As the sun burns into the Pacific, the dunes turn the color of a ripe peach, then gradually shade into a bronze-pink alpenglow. When the light is nearly gone and the grays are nearly black, there occurs a convincing illusion that the dunes are actually a monumental, storm-ravaged sea frozen in mid-tumult. At such a time the dunes' bond with the sea seems powerful and close.

In the concealed world of the sandy beach, one group of animals lives with consummate openness, and that is the birds. Freed from bondage to the ocean by their wings, they curve along with the sea breeze or scamper about at the edge of the surf with insouciance.

Often the first birds you see on a beach are sanderlings, which occur on almost every sandy

Constantly scampering just ahead of the tide, sprightly sanderlings refuel by snatching beach hoppers and food that washes in from the sea.

beach in the world. Unfortunately, unless you're an accomplished birder you cannot be positive that the half-foot-high, whitish-grayish-brownish bird is a sanderling, because many look-alike birds—knots, dunlins, stints, and sandpipers —also roam the shore. Birders get headaches from flipping through their field guides in futile attempts to sort them out, but anyone can get hours of pleasure from watching them race up the beach one mincing step ahead of a wave. Their skinny legs whirring, the little birds always

seem shocked to see a wave coming, though it has happened 10,000 times before.

Shorebirds provide lessons as well as comedy. Sanderlings and willets, for example, demonstrate how well defined the niches of a sandy beach can be. Though the species may be feeding only yards apart, the larger willets will forage closer to the water, where they eat adult mole crabs, and the smaller sanderlings will scurry farther up the beach, where they eat young mole crabs.

Sometimes as a black turnstone walks down a beach flipping over piles of algae to get at food, a sanderling will trail in its wake. After the turnstone has picked a few delicacies, the sanderling steps up to probe for morsels that turnstones don't eat. A sanderling may follow a turnstone for as long as half an hour, driving off other sanderlings that try to horn in.

Far less common than the sanderling is the snowy plover. Though snowies are doing all right at several lakes in the interior of the country, the coastal population has shrunk to a few hundred. They're listed as endangered in Washington, a species of special concern in California, and threatened in Oregon.

European beach grass was introduced to stabilize the dunes, but it spread so uncontrollably that it has reduced the number of good nesting sites for plovers; floods and storms also wipe out the nests.

Snowy plovers favor nesting sites on open sand, where their white and gray plumage effectively camouflages them from natural predators. But camouflage and their silent ways don't protect them from humans. One study on Monterey Bay revealed that people destroyed 14 percent of the snowy plover nests by crushing them under the wheels of off-road vehicles, by stepping on them, or by taking the eggs.

The plight of the snowy plovers along the Pacific Coast dramatically illustrates the impor-

tance of perceiving the sandy beach as a natural system and not just as a playground. Partly it's a problem of visibility; majestic conifer forests and animals like elk and wolves evoke concern much more readily than do unseen clams and plankton. But even though the life of the sandy beach may be largely hidden from our eyes, we can't allow it to be hidden from our minds or from our public policy, because it can't hide from our influence. 🐚

A winter resident of the coastal beaches, the marbled godwit (above) migrates elsewhere to nest. The snowy plover (left) stays year-round, its light colored feathers camouflaging the nest it scoops out in the 'sand.

aves dominate the coast. They steal sandy beaches overnight and break down rocky shores over years. Their perpetual motion seizes the viewer's eye and their muffled crash monopolizes the listener's ear. I remember the scorn I felt when I first visited America's Gulf Coast and saw the puny waves. (Obviously, I wasn't there during a hurricane.) Having grown up around husky Pacific breakers, I just couldn't regard a shore with 98-pound-weakling waves as a legitimate coast.

But I was being narrow-minded. Gulfs are an integral part of the coastal fabric, as are bays, bights, sloughs, straits, estuaries, coves, sounds, inlets, lagoons, and other places that are protected from waves. The coast wouldn't be as fascinating without these variations on the open beach. To whatever degree they are shielded from the force of waves and weather, these sheltered sanctuaries provide habitat for animals and plants that aren't found on the exposed coast.

Human beings are among the creatures that seek out safe harbors. Yet some of the measures humans take to allow ocean-

Protected coastlines such as Monterey Bay (left) provide a feeding ground for gulls, cormorants and sandpipers (right); lush habitat for clams and oysters; and a recreation bonanza for humans.

Although oozing mud flats and squishy marshes might not be the most glamorous of coastal habitats, they are crucial to many shorebirds, such as the black-necked stilt, that probe the grassy shallows for food.

Not surprisingly, virtually all major coastal cities (San Francisco, Los Angeles, San Diego, and Seattle on the Pacific Coast, for instance) grew up around natural bays. Smaller coastal towns, too, have tended to sprout in the natural nooks and crannies of the coastline. The growth of marinas, wharves, and shoreline buildings mimics the colonies of marine organisms that thrive in protected waters.

At one end of the spectrum of protected coastal areas lie barely shielded places like California's Monterey Bay, a half-circle that, according to Jim Covell of the Monterey Bay Aquarium, "hardly deserves to be called a bay." That may explain why Monterey is a town of 30,000 and not a city of three million, despite the head start it got when it caught the eye of Spanish explorers as early as 1542. But the relative lack of growth in the Monterey area is a blessing, because it has allowed Monterey Bay to remain rich and beautiful. Besides, if the coastline of Monterey Bay hadn't been so lavishly endowed with life, it might not have attracted Ed Ricketts, and the devotees of the Pacific Coast would have been much the poorer.

Who was Ed Ricketts? You may already know part of the answer: Ricketts was the person upon whom John Steinbeck based "Doc," the main character in Steinbeck's famous novel *Cannery Row*. By most accounts, Ricketts bore a close resemblance to the freethinking, painfully honest, kind, and complicated soul depicted in the novel. Though *Cannery Row* makes it clear that "Doc" is the proprietor of a biological laboratory on Cannery Row, you wouldn't know from the book that Ricketts was a pioneering, largely self-taught marine biologist who wrote the lion's share of the classic book on Pacific intertidal life, *Between Pacific Tides*. Ricketts also co-authored, with Steinbeck, the book *Sea of*

front development—throwing up breakwaters, jetties, and sea walls—destroy the very beaches that spurred development in the first place. Like Tinkertoys set in front of a steamroller, these efforts in the long term are doomed. In one absurd episode, ten weeks after Ocean City, New Jersey, unveiled its new $5.5 million beach, storm waves washed it away. Eighty thousand dollars a day for a beach. Maybe the city fathers should have just flown everyone to the Bahamas.

Cortez, a wide-ranging account of their 1940 voyage to collect invertebrates along the coast of Mexico's Sea of Cortez. As valuable as the scientific material in these books is, what stands out are the personal observations, humorous and philosophical, and the sense of seeing the universe in the quiet water of a tidepool.

The bay that lapped at the back of Ricketts' laboratory is still something to see, though civilization threatens it. Its abundant life stems largely from the presence just offshore of Monterey Canyon, an underwater rift vast enough to engulf the Grand Canyon. From the depths of Monterey Canyon, a summer upwelling of nutrient-laden water nourishes all sorts of wildlife. Without donning scuba gear or setting foot in a boat, you can see several species of whales and dolphins, seals and sea lions, and many bird species, including shearwaters, terns, fulmars, cormorants, and the diving brown pelicans.

Ringing the Monterey Peninsula, sometimes just a few yards from shore, is the kelp forest, an undulating world of giant brown algae. Up to 125 feet tall—in spring they can grow up to two feet a day—these thick strands of kelp offer food and protection to hundreds of species. Only divers get to see many of these creatures, but anyone strolling the shoreline likely will spot the kelp forest's most celebrated resident: the California or southern sea otter.

On a recent winter morning I accompanied Michelle Staedler on her rounds along the coast of the Monterey Peninsula. Staedler, a biologist at the Monterey Bay Aquarium, was looking for tagged sea otters as part of an aquarium study. Finding sea otters, tagged and otherwise, turned out to be easy; in half a day we spotted a couple of dozen.

One of the first otters we saw was a female that a few weeks earlier had been released from the aquarium. Having been washed up on the beach, and suffering from anemia, she had been taken there for rehabilitation. The sea otter looked healthy and content as she rolled herself up in the kelp to snooze. Kelp beds literally *are* beds to sea otters. Instead of hauling out on rocks or on the beach, they bundle up in the kelp canopy, where they can drift off to sleep without drifting off in the ocean swells.

Staedler and I also saw many sea otters feeding, which is hardly surprising because they must eat voraciously to stoke their high metabolisms. Several were using tools in classic sea otter fashion. Floating on its back, tool (usually a rock) on its belly, a sea otter will clutch an urchin, crab, clam or other resistant food item and hammer it until the shell breaks. Staedler said she saw one sea otter that seemed to be taking advantage of the tide as a tool. With its forepaws, the otter grasped a mussel attached to a rock, then hung on tight in the rush of the receding wave until the force of the outgoing water pulled otter and mussel off the rock.

Several times that morning we watched mothers and pups. Sea otter mothers bear and raise only one pup at a time, but the demanding relationship keeps them busy. For the first eight to ten months of its life the pup depends on the mother for everything. The mother nurses the pup, grooms it, brings it solid food, and totes it around on her back or chest, all the while teaching it the many survival skills a sea otter needs. Seeing the mothers and pups and listening to their high-pitched squeals was heartening. Not too long ago it appeared that the world would never enjoy another generation of sea otters.

In the early 1900s sea otters were nearly extinct. Furriers prized the luxuriously dense fur of the sea otter even above sable and mink,

By entwining themselves in kelp, a sea otter and her pup can doze without drifting off in the currents. Otters even ride out fierce winter storms wrapped in the streamers of this seaweed that can grow to be 100 feet long.

so hunters slaughtered sea otters by the hundreds of thousands. Though it seemed to be too late when the sea otter finally received official protection in 1911, the survivors eventually staged a comeback. Today the northern sea otter population in Alaska is thriving (estimates range from 100,000 to 200,000).

The southern sea otter population is cause for worry. Some of the 1,800 to 1,900 sea otters that live off the central California coast drown in fishing nets; others are shot by angry fishermen who feel that sea otters threaten their livelihoods. What really strikes fear in the hearts of otter lovers is the threat of an oil spill. A major spill could devastate the 225-mile range of the southern sea otter. And with more than 350 tankers passing along here each year and relentless pressure to permit offshore drilling, the danger is all too real. In February, 1990, for example, 290,000 gallons of Alaskan crude spilled out of a gash in a tanker near Huntington Beach, threatening wetlands and wildlife.

California sea otters and all the life in Monterey Bay will benefit from its designation early in 1990 as a National Marine Sanctuary. Nature

protected such bays from the full force of the sea, but only humans can protect them from the full force of humankind. To judge the need for protection, just look a few miles north to beleaguered San Francisco Bay.

San Francisco Bay was born with all the advantages. Virtually landlocked, exposed to the sea only through the mile-wide mouth of the Golden Gate, it lies at the other end of the spectrum of protected areas from wide-open Monterey Bay. Into the blessed San Francisco Bay poured many rivers and more than a hundred creeks, which created a wonderful variety of habitats. An early visitor, George C. Yount, wrote in 1833: "The wild geese, and every species of waterfowl darkened the surface of every bay.... The rivers were literally crowded with salmon.... It was literally a land of plenty." This sounds like romantic exaggeration, but other records support the substance of Yount's account, reporting whales so plentiful that spouts blew every half minute, and flocks of ducks so dense that several would fall to a single shell.

Today the waterfowl and salmon in the bay are greatly depleted. Gone also are the hundreds of Indian villages whose inhabitants lived the good life on the once-bountiful bay. In their place are about six million urbanites and a vast asphalt jungle. About half of the bay's open water and tidal marshes have been filled or cut off

Vast forests of kelp dissipate the force of the surf, providing protected habitat for sea urchins, snails, kelp crabs and fish. These quieter waters have even served as a refuge from storms for passing sailboats.

Like San Francisco Bay's East Brother Light Station, many Pacific Coast beacons have been automated and their dwellings restored as inns. Many are in places where nature-lovers can view abundant wildlife.

by levees and dikes, ruining 95 percent of the historic wetlands.

Perhaps more destructive in the end than the visible development will be the pollution. As naturalist John Steiner of the San Francisco Bay National Wildlife Refuge said: "The bay is being assaulted from all sides. It has toxins galore."

About the only manmade intrusion into this once-paradisiacal bay that seems to belong is the East Brother Light Station. Built in 1873-74, this Victorian beauty stands on one-acre East Brother Island. The Coast Guard automated the light and foghorn decades ago, but a nonprofit group has restored the station—without taint-

ing authenticity—and runs it as an elegant bed-and-breakfast inn. After an evening stroll to look at the lights of San Francisco and the seabirds, guests sit down to a five-course candlelight dinner, then retire to rooms furnished with antiques. The former keepers of the lighthouse surely didn't live so opulently, but neither did they pay $285 a night.

East Brother Light Station is a pleasant oasis, but what of the rest of the bay? Is there any hope for San Francisco Bay? That was the question I put to John Steiner one afternoon as we drove around the wildlife refuge that was established to protect some of the sensitive south bay from further development. The question was inevitable during stop-and-go traffic en route from one parcel of the refuge to another.

Steiner's reply boiled down to this: the paradise that prompted Yount's purple prose in 1833 never will return, but measures can be and are being taken to preserve pockets of wildness.

Steiner showed me some former salt evaporation ponds whose levees had been breached the previous year. Already picklegrass had appeared and thousands of shorebirds were picking it over for food. Later Steiner pointed out a long stretch of shoreline that used to be a Nike missile base and is now returning to a wild state. He said that tens of thousands of acres in the San Francisco Bay area have been saved from development, incongruously enough, due to their use by the military. Scared by the excesses of the past, the citizens question all development, but Steiner fears that their resistance will weaken as population pressures intensify.

Despite the loss of so much habitat, the bay still accounts for 90 percent of California's remaining coastal wetlands, which makes it vital to birds. About 70 percent of the millions of birds migrating along the Pacific flyway at least pause there. The enormous numbers and variety of waterfowl and shorebirds excite the most admiration—and concern.

As we drove around the bay, we saw a knot of avocets here, a flock of huge white pelicans there, and sandpipers seemed to be everywhere. But so were the signs of people—power lines, bridges, highways, railroad tracks, levees, old mines, salt ponds, smog, industrial parks, and all the other manifestations of civilization.

To see a bay in its more pristine state, however, one needs only to look several hundred miles to the north. Shorebirds can be found twittering by the thousands at Grays Harbor on the Washington coast—one of the most spectacular gatherings of shorebirds in North America. When conditions are just right each April, half a million migrating shorebirds crowd like Super Bowl fans into Bowerman Basin, a tidal flat on the bay's north end.

Birds don't congregate in such protected coastal areas by accident. They come for food, and it's quite a buffet. These sheltered waters are tremendously productive. Estuaries and their associated marshes and mud flats historically have been thought of as wastelands, largely because their wealth is mostly microscopic or below ground. Rich Everett, a marine biologist who studies estuaries, said, "It's not like going out on the Serengeti. Everything here is smaller, and you have to focus on that smaller world."

Though the worms, clams, snails, shrimp, oysters, crabs, whelks, and sea cucumbers lack the charisma of avocets and peregrine falcons, they're intriguing if you can adjust your perspective. If you squelch out onto a mud flat after the tide has gone out, the first thing you'll notice is the evidence of the bustling city beneath your boots: mounds, holes, depressions, tiny pools, and fecal pellets. One resident, the pink ghost shrimp, is such a prodigious burrower that you must be wary of their extensive beds, or else you may sink into muck up to your waist. Scientists who study these shrimp wear snowshoes.

If a mud flat looks like just a bunch of mud to you, pick up a handful and take potluck. The odds are with you. It has been estimated that a handful of mud taken from a tidal flat in San Francisco Bay will, on average, contain 20,000 living creatures. Most of those creatures are tiny,

of course, but occasionally your shovel will strike pay dirt, unearthing a larger animal, such as a moon snail. The shells of these mollusks measure several inches across and their foot (the organ used for crawling and smothering the snail's victims) is several times larger than the shell. Moon snails often will be seen partially buried in the mud or as a lump just beneath the mud's surface, cruising in search of prey at a stately inch per 15 seconds.

Another resident of the mud flats, the fat innkeeper worm, employs a more elaborate feeding method. First it constructs a U-shaped burrow about a foot deep with entrances up to a yard apart. Next the innkeeper spins a slime net, attaching one end to an entrance and playing out the rest of the net while moving down into the burrow. Then the innkeeper begins rhythmically contracting its white, sausagelike body.

While permitting the passage of water, the incredibly fine net traps food particles less than a millionth of a meter in diameter. After about an hour the net will be clogged with food, and the innkeeper will move up the burrow swallowing the net—and the food—along the way. This skillful worm is called an "innkeeper" because pea crabs, a scale worm, and little fish called gobies almost always share its burrow, free-loaders enjoying the food, protection, and oxygen that the inn provides.

Such intriguing animals underscore the complexity of the coast and make a sojourn there all the more enjoyable. Still, only the staunchest student of invertebrates can keep his attention focused on worms and snails when shorebirds are present.

Shorebirds are arguably the most beautiful group of birds in America. Certain species, such as black-necked stilts and Wilson's phalaropes,

Migrating birds such as these sandpipers funnel through Bowerman Basin, Washington, every spring. Vital resting and refueling stops, such sanctuaries supply vast quantities of food just when birds need it most.

If a predator threatens, the moon snail can eject large amounts of water from its foot in order to contract inside its shell. When the body is fully expanded, usually as the snail moves around, it can engulf the shell.

embody all the ethereal qualities that evoke the essence of birds. Take the American avocet: elegantly long, blue-gray legs; well-proportioned black-and-white wings and body; a soft, bronzed ballerina's neck; and a long, slender, delicately upturned bill. The avocet is a study in grace, right down to its mellifluous name.

Though in far lesser numbers than shorebirds, raptors, too, seek out protected pockets along the Pacific Coast. Bald eagles cast shadows on bay waters as they hunt for fish. Northern harriers reconnoiter marshes in search of small

A peregrine falcon (left) rests after catching its prey. While hunting, it sometimes plunges at 180 mph, dive-bombing smaller birds. If a hawk or falcon threatens, dunlins (far left) take off in a tight cluster, swirling and twisting to confuse the predator.

mammals. Ospreys patiently circle, waiting for just the right moment to divebomb from the sky after a fish. And then there are those raptors such as merlins and peregrine falcons that show a taste for shorebirds.

Given the peregrine falcon's unsurpassed prowess at taking birds, and given the masses of shorebirds to be taken, you'd expect a peregrine could grab a shorebird as easily as we grab a barbequed chicken leg from a platter, but such is not the case. The immense numbers of the shorebirds actually work to their advantage, because a peregrine strikes most effectively when it can zero in on a single bird.

I once witnessed this phenomenon at Point Reyes National Seashore in California. A flock of sandpipers was spooked by an incoming peregrine and darted off in a tight formation that would have made the Blue Angels envious. The sandpipers flew quickly, but that had nothing to do with their salvation; the streaking peregrine overtook them with ease. However, as the falcon closed on the sandpipers, it slowed, apparently confused. Throttling down to match the speed of the sandpipers, it waded into their midst but couldn't decide which to seize, like a child unable to choose a piece of candy from a full box. The peregrine started in one direction, then reversed itself, then seemed to stall out as the sandpipers scattered in all directions.

In a few seconds the assault was over, and the falcon winged away empty-taloned, climbing into the gray sky until it blended into a background of clouds the color of the winter sea below them. The peregrine may have missed one opportunity, but no doubt it would find others as it scouted the lagoons, bays and other bountiful backwaters of the coast.

The power of the Pacific Ocean is difficult to overlook. Even clothing it in a balmy summer day deceives no one; that's like trying to hide Arnold Schwarzenegger's muscles with a tight t-shirt. Waves build over the ocean's 64 million square miles—more than twice the area of its nearest rival, the Atlantic—and move toward the coast like rolling hills. Set aside the anomalies, such as huge waves called tsunamis that are created by underwater earthquakes or volcanoes. In the misnamed Pacific, even plain old wind-generated storm waves commonly reach heights of 25 and 30 feet. On occasion they loom higher than 60 feet. While the waves tend to obliterate any measuring equipment placed in their path, no one could argue with the verified account of a 100-pound rock that Pacific waves threw through a lighthouse window 195 feet high.

Any swimmer who has been tumbled in a plunging wave has been touched by the power of the Pacific, and even an idyllic-looking sandy shore holds unseen dangers. A steep, narrow beach with coarse sand—the kind formed by heavy wave action—is particularly deadly. The waves that move in around the outlying rocks catch children and even adults offguard in the shifting sand, and then suddenly sweep them out to sea.

To witness the full might of the Pacific, try making a pilgrimage to the rocky shore during a storm. The rocky shores stand toe-to-toe with the sea, taking punch after punch in a titanic heavy-

How animals cope on a wave-wracked shore is a marvel. Some burrow in the sand, others cling stubbornly to rocks. Birds such as the western gull (right) comb the coast for sea stars and other bounty from the sea.

weight bout. I had ringside seats for these championship fights when I lived in the coastal town of Santa Cruz, California. Sometimes during storms I walked down at night to the bluffs just north of the lighthouse and—though I knew my vantage point was a dangerous one—watched the sea hurl itself against the land.

One night as I stared into the rumbling darkness, one of the undulating shadows detached itself from the blackness and steamrolled toward shore. By the time the wave was near enough to be seen distinctly, it was as high as a two-story house. About 50 yards from shore the wave reared up and prepared to break, the leading edge of the wave fracturing into writhing foam. Like a bighorn ram lowering its head right before butting its rival, the wave curled under just as it slammed into the undercut bluff, and watery shrapnel exploded over the bluff and over me and even over the road.

A split second after hitting the brow of the bluff, the wave ended its journey by thudding against the bottom of the undercut bluff, 40 feet below and 30 feet *behind* where I stood. I couldn't see this body blow, but I could feel the earth buckle, and I knew that the bluff had been undercut a little more. In the years since, several of the bluffs I frequented as a youth have collapsed and their remains have been dragged into the Pacific's depths.

People, as well, get dragged into the ocean all too often. I was once swept off a rocky point by a sneaker wave—a gigantic wave that rolls ashore out of sync with the others—so I can attest to the danger posed by the sea. The need for caution extends to sandy beaches, too. I witnessed another surprise-attack wave one sunny day on a flat southern California swimming beach. Thousands of sunbathers were rudely surprised by two feet of rushing water that left a flood of overturned picnic baskets, beach umbrellas, and coolers in its wake. If that wave had surged up a narrow sandy beach backed by steep cliffs, the result might have been fatal.

It's a good idea to spend a few minutes watching an area to see where the waves are breaking and how big they are. Be especially wary of beaches backed by steep cliffs, or of a beach hemmed in by bluffs you cannot climb.

Solitary walks on the beach can be memorable, but if you want to be around to remember them, take someone along when you're heading for the intertidal area. Experts advise going on walks during an *ebbing* tide. On an *incoming* tide, the tide can sneak in behind you and cut you off from land. The shore, especially the rocky shore, is not to be trifled with.

Sailors have known that since boats first put out to sea. Many ships have been reduced to kindling and scrap metal on the unforgiving rocks of America's Pacific Coast. One of the most treacherous stretches runs from Cape Flattery at

the tip of Washington's Olympic Peninsula south for more than 100 miles, a stretch devoid of natural quiet backwaters or safe harbors. In addition to the many typical shipwrecks, this coast has been the site of exotic wrecks brought about by the Kuroshio Current off the Japanese coast.

In 1927, a fishing junk named the *Ryo Yei Maru* was sighted just south of Cape Flattery and taken in tow to port. Aboard were human bones, decaying bodies, and the ship's log, which told a grim tale. The engine failed on December 12, 1926, when the ship was about 700 miles from Japan, and the junk drifted in the Kuroshio Current, then farther into the Pacific Ocean, for the next several months. The log's May 10 entry said it all: "Only Captain Miki and I remain alive. Both . . . too weak to tend the helm."

Monster waves, collapsing bluffs, shipwrecks, ghost ships—the rocky shore can be a rough neighborhood to pass through. Imagine *living* there. A lot of plants and animals *do* live there, adapting and thriving among those rocks that get mauled by the surf.

The rocky intertidal area is that part of a rocky coast which is submerged at high tide and exposed at low tide. With two high tides and two low tides every day, this state of flux makes the rocky intertidal zone a harsh realm. Inhabitants must contend with exposure to sun, wind, air, predators, and scouring tidal flows. Yet animals and plants that can cope reap many benefits, enjoying twice-a-day helpings of the rich broth of the ocean and the high oxygen content of the churning waters.

Rocky intertidal areas beckon curious visitors along much of the Pacific Coast, but none can compete with the wild stretches of Olympic National Park's 57-mile coastline in Washington.

Green anemones get their color partly from algae that live in their tissues. Some anemones can reproduce all by themselves (left), stretching until they divide, a process that takes about two days.

Hike there at low tide and you're likely to come across a landscape of fractured shale reef, boulders, and tidepools, some a riot of iridescent algae, others so still and lucid that the water between you and the creatures on the bottom seems to vanish.

If you arrive at one of the park's rocky intertidal areas at high tide and nothing much is visible, be patient. The sea has its own way of doing things. Take a look at the rocks that get flooded only during the highest tides, and even there you may spot tiny periwinkles and some limpets. Follow the receding tide down toward the nurturing ocean, and you'll find abundance within a few steps, as if you'd walked from the desert into a lush, irrigated field.

I remember *hearing* the intertidal zone's abundance one dark morning when I went tidepooling at Cape Alava, near the northern end of Olympic National Park. I started out before dawn, unable to see beyond the narrow confines of my flashlight's beam. After a few steps on the rocks I began hearing things—little scraping, scuttling, splashing sounds. I stopped, and the sounds stopped. I started, and the sounds started. I had an eerie feeling that the rocks were alive, whispering. Slowly, I shined my light around and saw thousands of black turban snails. But surely those stolid mollusks

Mixed in with the hermit crabs under the rock were eels, brittle stars, lined shore crabs, aqua porcelain crabs, and snails. And that was just a modest rock. When I was finished looking I carefully lowered the rock to its original position. The cardinal rule of tidepooling is to leave things as you found them. If you replace a rock sloppily, pry an animal off a rock, or move a creature from its home to an inappropriate habitat—sometimes only a matter of inches—you may be condemning it to death.

That such a mob lives beneath a rock suggests an underlying principle of intertidal life; these creatures owe their first allegiance to the sea, not to the land. Those basement-dwellers were seeking moisture and conserving moisture until the life-giving tide returned.

Other animals survive by wedging into cracks in the rocks, by staying in tidepools, by having a sticky body surface that retains a protective coating of sand, or by holding moisture by shutting their operculum, a little hatch inside their shell. Sitting high on the rocks, where it is only splashed by waves, the eroded periwinkle can live out of water for up to two months because it seals in the moisture from even an occasional splash of water.

The biological riches of the intertidal area increase as you follow the receding tide. Often you can discern distinct bands where one group of species yields to the next, zones that are determined by the extent of tidal exposure: the splash zone, the high zone, the middle zone, and the low zone.

The vast congregations of black turban snails characterize the high intertidal zone; in the middle zone, even denser assemblages of California mussels can be found. Packed shell-to-shell to withstand the surf, mussels often coat

Sea stars haunt mussel beds to look for a meal. Latching onto an unlucky victim with its strong, suction-cup feet, a sea star pulls until the prey's shell opens enough to get at the tasty morsel inside.

weren't the source of the sounds? Baffled, I played the flashlight quickly over some nearby rocks. This time I realized that I'd been hearing not snails but skittish hermit crabs, hundreds and thousands of them that were merely occupying the turban snail shells for a while.

To get a better look at the elusive hermit crabs, I turned over a rock. There they were in their snail shells, always renting, never owning.

the rocks like black paint. In northern California at Duxbury Reef, single beds of mussels number two and three million, and untold numbers inhabit the acres of shale reef that emerge at low tide. No wonder coastal-dwelling peoples from prehistoric days to the present have gathered these shellfish for food.

However, then as now, paralytic shellfish poisoning (PSP) can make mussels a delicacy literally to die for. A hundred or more Aleut Indians working for Aleksandr Baranov, the first governor of Russian Alaska, died from a feast of toxic mussels. It would be wise to heed the quarantines on mussels that often go into effect during the summer, and content yourself with merely looking at mussels.

Especially worth seeing are the byssal hairs, strong fibers that anchor mussels to a rock and look like fishing line. Find a gap in a mussel bed and you can feel the fibers, created when the mussels extrude a fluid that hardens. And does it harden. I once tried to pull apart a length of byssal hair and only succeeded in ripping my hands. Byssal hairs are so sturdy that they have attracted the attention of researchers seeking the perfect underwater cement.

California mussels would exist in the lower intertidal area as well as the middle zone were it

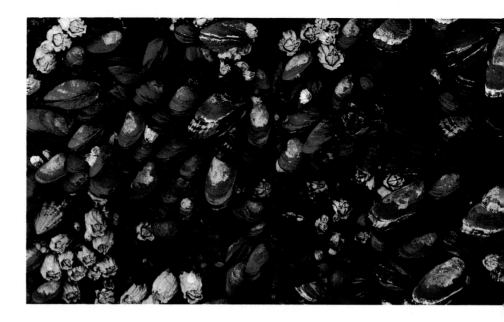

not for the ochre star, the most common sea star (starfish) on the Pacific Coast. Ochre stars —which also come in purple, tan, brown, and orange—crave mussels even more than we humans do, and a single ochre star can consume 80 mussels a year. A voracious sea star will crawl atop a mussel, evert its stomach into the slightly open shell, and extract the mussel.

To escape this grisly fate, some intertidal animals, such as limpets and snails, flee when they detect the "scent" of an approaching ochre star. Mussels, however, are tied down. On a recent wander through Olympic National Park's Starfish Point I saw dozens of sea stars draped over doomed mussels that had unwisely made their homes below their safety zone.

Ochre stars typically are an intertidal visitor's favorite animal, but when a sunflower star eases into view, all other intertidal creatures seem to grow dim. Its size alone—commonly at least two feet across—makes it commanding.

Even if they weren't enormous, sunflower stars would catch your eye with their colors—a riot of red, purple, yellow or orange. Presumably they were named because of their sunflower shape, but the name might have come

Barnacles and blue mussels (above) have adapted to an unstable environment by firmly cementing themselves to rocks. Constantly in search of bigger homes, hermit crabs (left) move from one snail shell to another as they grow. The crab's flexible body curls to fit each new shell.

from the varieties with a striking, deep orange color, a hue that brings to mind the setting sun. Even more radiant are the slashes of fiery orange that flash from among the folds of the sunflower star's soft skin, like streams of lava burning in cracks in the earth.

Finally, sunflower stars wow onlookers because they move along on 20 to 24 rays (legs), and they move fast. They are the cheetahs of the Pacific sea stars, bounding along at 20 inches a minute. By contrast, an ochre star crawls about 1 inch a minute, and that's when it is feeling frisky.

Like many of the world's 1,600 species of sea stars, sunflower stars can regenerate lost rays, but they take this ability one step further: they can shed a ray at will. When pincered by a king crab, for example, a sunflower star can drop the clamped ray and dash away. Of course, such evasive tactics aren't often necessary, because few creatures prey on the sunflower star.

The sunflower star preys on many; however, its food of choice is the sea urchin, which it generally swallows spines and all. This taste for sea urchins—minus the spines—is shared by many humans, though in America only patrons of sushi bars and a sprinkling of Portuguese-, Italian-, and Samoan-Americans favor urchins. Many Japanese consider sea urchins a delicacy, and their liking for urchins has led to a boom in sea urchin harvesting along the Pacific Coast.

The gold sought by marine miners is the yellow gonads of the red sea urchin. Called "roe" by those in the industry and "uni" by the Japanese, the gonads by any name fetch up to $30 a pound retail. The industry got started in the early 1970s in southern California, where the harvest has plateaued at about 20 million pounds of unprocessed sea urchins annually.

In northern California, on the other hand, a gold rush atmosphere still prevails. Virtually no urchins were taken as late as 1984, but 30 million pounds were harvested in 1988. The sluggish economies of small coastal towns like Bodega Bay, Point Arena, and Fort Bragg have been boosted substantially by this bonanza. Divers working in the shallow waters offshore are making up to $30,000 a year. And you can bet processors and middlemen make their share.

But northern California's sea urchin boom is rushing toward a bust. After only four years of

The largest sea star on America's Pacific Coast, the sunflower star (above) has at least 20 rays when fully grown and often stretches two feet in diameter. Abalones (below) were once harvested by the ton, but taking them commercially is now strictly limited.

boom the fishery is showing signs of decline because of overharvesting. Over the decades, unusually high stocks of urchins built up because sea otters—predators on urchins—had been driven from these waters by fur traders. Sea otters had kept the red urchin population density at about 1.5 percent of what it is now, but once the sea otters were gone, dense beds of urchins developed. Now they are thinning out.

Biologists, processors and enlightened divers are pressing for regulations to make the urchin fishery sustainable, but so far the California legislature, which exercises greater direct control over fisheries than is the case in most states, seems reluctant to act.

Over the years another resident of the low intertidal area, the abalone, has spawned even more commercial interest than the Johnny-come-lately sea urchin. The quest for the delectable meat of this enormous snail almost led to the founding of an independent island nation off the southern California coast that was to have been called "Abalonia."

In 1966 a group of southern California businessmen concocted a scheme that called for a freighter to be scuttled on the abalone-rich Cortes Bank, about 100 miles west of San Diego and (according to the businessmen) outside U.S. territorial waters. They planned to declare the freighter, which would protrude above the sea on the shallow bank, a sovereign country. Then they would build an island roughly a mile in diameter and mine abalone to their hearts' —and bank accounts'—content.

The future rulers of Abalonia bought a freighter named the *Jalisco* and towed it out to the bank. But before they could get clearance to scuttle the *Jalisco,* the sea did it for them, and in the wrong place. Massive waves rose suddenly and overwhelmed the freighter, and the four would-be Abalonians were swept overboard.

The red sea urchin's large size sets it apart from others of its kind. The smaller purple variety digs in cavities in rocks to withstand the waves.

163

Immediately we understood why nudibranchs have been called the "butterflies of the sea." Entrancing. The first one we saw was no garden-variety slug. Through our loupes it looked like a luminous, translucent vacuum cleaner made of liquid opal. From its back grew a riot of golden poppies shimmying in the water. Gaily colored, mushy, fluttery, *Hermissenda* looked like a softie, but it's one of the toughest customers in the intertidal zone. It is a feared predator that nibbles with impunity on sea anemones and hydroids.

Anemones and hydroids bristle with stinging cells that deter most predators. However *Hermissenda* somehow ingests these cells without triggering the sting and passes them to the tips of its own gills to be used as defensive weapons. If you think this is easy, consider the case of an impetuous researcher who, in headlong pursuit of knowledge, kissed a sea anemone. Knowledge has its price. The researcher's lips swelled into hot, red balloons and landed him in the hospital.

Leaf barnacles (above) frequently attach themselves in great masses to wave-swept rocky cliffs. With the aid of a shell-dissolving enzyme, these whelk snails (below right) drill holes into the shells of barnacles and mussels and begin their feast.

Only heroic action on the part of the tugboat crew that had towed the freighter saved the four men. When last seen, Abalonia was awash and listing at 30 degrees.

With Abalonia gone to join Atlantis and the northern California sea urchin fishery heading for a fall, perhaps we should turn our attention from commercial ventures to intertidal dwellers that are a pure, unexploited joy—which brings us to nudibranchs *(NEW dub branks)*.

Bob Stewart, a naturalist for Marin County in California, introduced me and ten other tidepoolers to nudibranchs a few years back during an exploration of the rocky intertidal zone at Duxbury Reef. Stewart's excitement about these sea mollusks finally pulled us from the sea stars and giant green anemones over to the tidepools dense with seaweed. We weren't enthralled with searching for half-inch-long sea slugs in the icy water and slimy seaweed—not until we found one, that is.

Though the aptly named brittle stars (left) are fragile, they can generate a new arm if one is lost.

Next page: Undaunted by creatures that sting, nudibranchs such as this bright orange one safely feed on these animals. Then they transfer the poisonous cells intact to their own tentacles for use in their own defense.

Now, nearly as excited about nudibranchs as Stewart was, we sifted through seaweed for much of the morning, eventually turning up 17 more species of nudibranchs. Among them was a Hopkins' rose, which looked like a hot-pink feather duster; a chalk-lined dirona, which looked like a minuscule mountain range of snow-frosted pinnacles; and a brilliantly orange red-sponge nudibranch, which lay camouflaged on a sponge of the same color. We also encountered spiky purple sea urchins, intricately marked chitons set on the rocks like gems, and fields of fat leaf barnacles.

The incoming tide, however, always returns to cover the world of the sea urchins, the sunflower stars, and the nudibranchs. A beachcomber could walk by at high tide and never know that beneath that veil of water, millions of animals were hunting, eating, mating, giving birth, fighting, and dying.

If you've been peering into the Lilliputian world of damp seaweed thickets and rock crevices, you will need to readjust your perspective when the rising tide covers the last black turban snail and the spell of the intertidal area is broken. It's something like emerging from a movie theater after a matinee and blinking at the unreality of the daylight world.

Take a look at the sea stacks, for example. Like the blocks of some watery Stonehenge, these rock monoliths jut from waters near many rocky shores. The heavy hammer of the Pacific sculpts sea stacks by chiseling away the softer materials of sea cliffs until only the harder rocks are left behind. The tops of sea stacks make safe nesting sites, and in the proper season they are thronged by pigeon guillemots, cormorants, murres, auklets, and other sea birds.

Other notable birds spend more time on shore leave. Two engaging species associated with the Pacific rocky shore are the American black oystercatcher and the black turnstone. Using their short, slightly upturned bills, black turnstones flip over surprisingly large rocks to get at food. Oystercatchers employ an even more skillful technique, slipping their long, flat bills into gaping oysters and other bivalves and severing their adductor muscles. When not performing surgery, these portly shorebirds mince around on pink legs, sometimes engaging other birds in games of peek-a-boo.

Mammals show up on the rocky shore, too, spending time on offshore rocks, on the beach, and even in the intertidal area. Harbor seals, for

example, while away hours dozing, draped over rocks like huge, limp sausages. Their neighbors on the rocks might be Steller sea lions; if so, you'll know it. Male Steller sea lions measure up to 12 feet long and weigh up to a ton, about five times the weight of a harbor seal. The name "sea lion" fits them perfectly, for they sport shaggy manes and roar like the devil.

Steller sea lions seem to have an aversion to being partly immersed in water—they prefer all or nothing—so rather than crawl into the sea, they dive, sometimes from high rocks. More improbably, they eschew crawling out of the sea and instead somehow launch their bulk from the water onto the tops of rocks. Some surprise for an unsuspecting tidepooler.

The rocky shore harbors many surprises for the patient and lucky observer, some as astonishing as the dramatic entrance of a Steller sea lion, some as easily overlooked as the miniature beauty of a nudibranch. Some of the rocky shore's surprises defy the pigeonholes into which we arrange our understanding of the world. One such serendipity came my way years ago on a remote Pacific beach.

I was walking along the shore when I spotted three sleeping sea lions sprawled across some rocks at the water's edge. Unaware at the time that sea lions are exceptionally fearful of people, I began sneaking up to this trio to get a close-up photo.

When I was about 50 feet away, one of the sea lions spied me. He pushed onto his flippers, swayed nervously, and plunged into the water. A second soon followed. But the third remained oblivious, flopped across the rocks like a hound dog snoozing in the sun. I crept closer and closer until I was only about five feet from him.

Startled by the noise of the camera's shutter release, the sea lion jerked upright and howled at me, gyrating madly as he apparently tried to figure out how to safely turn his back on me in order to dive into the water. Before he could resolve his dilemma, a strange impulse seized me—I began singing to him.

I have a voice that can only aspire to mediocrity, but somehow it came to me that singing might communicate my friendly intentions and my apology. At least, looking back, that's how I see my impulse. I settled on Brahms' *Lullaby*, making up lyrics as I went, singing such gems as, "Don't be alarmed, I mean you no harm" and "Please don't go, I'm not your foe." I assumed the sea lion would bolt any moment—if not from me, then from my voice—but he didn't. Instead, bit by bit, he relaxed, slowly settling back onto his rocky bed. After 20 minutes of my serenade he closed his eyes and went back to sleep. Feeling like laughing and crying at the same time, I tiptoed away, like a parent slipping away from a sleeping child.

A noisy hubbub fills the air along rocky ledges during the breeding season of both Steller and California sea lions (far left). These sea lions are well equipped for deep-sea diving (left), with insulated skin and a circulation system which is energy-efficient.

INDEX

PHOTO CREDITS

Library of Congress Cataloging-in-Publication Data

Treasures of the tide
National Wildlife Federation.
  p.   cm.
  ISBN 0-945051-21-2 (trade)
  ISBN 0-945051-22-0 (leather)
  1. Seashore ecology—United States. 2. Seashore ecology—United States—Pictorial works.
3. Coasts—United States. 4. Coasts—United States—Pictorial works.
  I. National Wildlife Federation.
QH104.T74   1990
574.5'2638'0973—dc20   90-5921
                        CIP

CREDITS

NATIONAL WILDLIFE FEDERATION

Jay D. Hair, *President and Chief Executive Officer*

William W. Howard Jr., *Executive Vice President and Chief Operating Officer*

Alric H. Clay, *Senior Vice President, Administration*

Francis A. DiCicco, *Vice President, Financial Affairs*

Lynn A. Greenwalt, *Vice President, International Affairs and Special Assistant to the President*

John W. Jensen, *Vice President, Development*

Alan B. Lamson, *Vice President, Promotional Activities*

Sharon L. Newsome, *Vice President, Resources Conservation*

Gary San Julian, *Vice President, Research & Education*

Larry J. Schweiger, *Vice President, Affiliate & Regional Programs*

Stephanie C. Sklar, *Vice President, Public Affairs*

Robert D. Strohm, *Vice President, Publications*

Joel T. Thomas, *General Counsel*

STAFF FOR THIS BOOK

Howard Robinson, *Editorial Director*

Elaine S. Furlow, *Senior Editor*

Donna Miller, *Design Director*

Bonnie Stutski, *Photo Editor*

Michele Morris, *Research Editor*

Cei Richardson, *Editorial Assistant*

Karen Stephens, *Editorial Secretary*

Paul Wirth, *Quality Control*

Margaret E. Wolf, *Permissions Editor*

Kathleen Furey, *Production Artist*

Design by Shub, Dirksen, Yates, & McAllister, Baltimore

SCIENTIFIC CONSULTANTS

Joel W. Hedgpeth, *author and emeritus professor of oceanography, Oregon State University, Corvallis, Oregon.*

Craig Phillips, *biologist and former director of National Aquarium, Washington, D.C.*

David White, *biology professor, Loyola University, New Orleans, Louisiana.*

NATIONAL WILDLIFE FEDERATION
1400 Sixteenth Street, N.W., Washington, D.C. 20036-2266

WRITERS

THOMAS B. ALLEN vacations on a rocky island in Maine every summer and lives in Maryland the rest of the year, frequently sailing on Chesapeake Bay. He has written for the National Geographic Society, the *New York Times Magazine,* and *Smithsonian* and contributed to the Federation's *Earth's Amazing Animals.* He is also the author or co-author of nine non-fiction books, including *Merchants of Treason,* a study of modern espionage.

BOB DEVINE grew up near the beaches of southern California and now lives in Oregon. An avid tidepooler, he wrote on the subject for *Travel & Leisure;* his work also has appeared in *National Geographic Traveler, Better Homes and Gardens, Mother Jones* and the *Boston Globe.* He was a contributor to *America's Wildlife Hideaways,* published by NWF Books. Formerly an associate editor for *Rocky Mountain Magazine,* he has been a freelance writer on nature topics since 1982.

DONALD DALE JACKSON writes regularly for *Smithsonian, Audubon,* and *Wilderness* magazines. Recently he has written on such varied subjects as gulls, nutria, and the exploration of America's western frontier. His books include *Gold Dust, Flying the Mail,* and *The Explorers.* He was a contributor to *America's Wildlife Hideaways,* published by NWF Books. Jackson lives in Connecticut, but says his favorite place along the Gulf Coast is the Louisiana bayou country.